A. Tagliamonte and I. Maremmani (eds.)

# Drug Addiction and Related Clinical Problems

Springer-Verlag Wien New York

Alessandro Tagliamonte, M.D.
Professor of Pharmacology
University of Siena, Institute of Pharmacology, Siena, Italy

Icro Maremmani, M.D.
Professor of Medicine and Drug Addiction
University of Pisa, Psychiatric Institute, Pisa, Italy



Typesetting: Camera ready by editors

Printed on acid-free and chlorine-free bleached paper

With 39 Figures

**Library of Congress Cataloging-in-Publication Data**

Drug addiction and related clinical problems / edited by Alessandro Tagliamonte and Icro Maremmani. p. cm. Includes bibliographical references. ISBN-13:978-3-211-82674-4 (alk. paper) 1. Substance abuse – Chemotherapy. 2. Substance abuse – Molecular aspects. 3. Methadone maintenance. 4. AIDS (Disease) – Prevention. I. Tagliamonte, Alessandro. II. Maremmani, Icro. [DNLM: 1. Substance Dependence – rehabilitation. 2. Methadone – therapeutic use. 3. Health Policy – Europe. WM 270 D79348 1995] RC564.D779 1995 616.86'061 – dc20 DNLM/DLC

ISBN-13:978-3-211-82674-4 e-ISBN-13:978-3-7091-6901-8
DOI: 10.1007/978-3-7091-6901-8

# PREFACE

Many of the questions related to the treatment of drug addiction originate from the confusion that still characterizes this issue, even though biology is irreversibly changing our outlook on the physiology and psychopathology of the mind. In fact, prejudices stemming from the archaic concept of mind-body dualism are so difficult to eradicate that even a skilled psychiatrist may find it hard to distinguish a psychiatric symptom from a socially transgressive behavior.

The dilemma becomes even more salient for drug addiction, since many of the substances which induce abuse and dependence are illegal, and their production, trade and use are forbidden by law. If the use of heroin is connoted as a crime, it may become controversial to recognize its chronic sequelae as a disease. Hence, withdrawal symptoms may alternatively be attributed to drug effects, or labelled as immoral attitudes arising from a vicious personality.

However, the physician's judgement, which has to be merely instrumental at improving the patient's quality of life, should never be influenced by such complex setting. In fact, the only concern of the physician must be that of ascertaining the causes of symptoms and of removing them: if symptoms are induced by the toxic effect of a drug, he has to use the most effective antidote, and then proceed towards a detoxification; if they are due to withdrawal, a substitutive substance should be given initially. In the case of narcotics, substitutive therapy on a maintenance basis may become mandatory. Treatment must be pragmatic, professional and never contaminated by non medical motivations.

In the present book the diverse issues related to drug addiction, from basic research to clinical treatments, are examined with a scientific approach, without ideological constraints.

In the first part, H.H. Loh discusses some of the most recent studies on opioid receptors c-DNA isolation and cloning, and the intriguing role played by OBCAM (opiate binding cell adhesion molecule) in the regulation of opioid receptor function.

The role of dopamine, and the interplay between different dopamine and opioid receptors in the mediation of rewarding and motivational effects of drugs of abuse in the limbic system are treated in separate chapters by A. Herz and G.L. Gessa. New relevant data indicating that endogenous morphine is the physiological ligand for μ-opioid receptors are presented by A. Tagliamonte.

V.P. Dole's chapter is an historical overview of methadone maintenance (M. M.)

therapy, which also points out that basic research plays a key role in understanding the psychopathology of drug addiction.

The sections concerning the medical aspects of drug addiction deal with:

1) the high prevalence of psychiatric co-morbidity among drug addicts (L. Daini);
2) the issue of polydrug abuse, sustained by the availability of recreational drugs like tobacco and alcohol, and by the excess of prescribed psychoactive drugs - such as hypnotics, tranquillizers, and stimulants (S. Rawaf). In particular, the problem of alcohol abuse, as well as that of cocaine, is especially considered in terms of complications in the therapy of heroin addicts (G.P. Guelfi, O. Zolesi).

Almost 80% of the clinical data deal with M. M. programs, and the main conclusions are:

a) the efficacy of M. M. in the treatment of heroin addiction has now been objectively demonstrated (M. Parrino, M. Rizzi);
b) the dose of methadone has to be tailored to each patient according to her/his individual and contingent needs;
c) the correct treatment with methadone reduces heroin use - up to its complete cessation in most patients, and helps to control the abuse of other psychotropic drugs;
d) from a medical point of view, it is irrational to discharge a patient from a M. M. program only because she/he abuses heroin or any other drug (R. Newman);
e) the possible complications associated with pregnancy, childbirth and infant development in drug dependent women can be markedly diminished if adequate care and pharmacotherapy (in the form of M. M.) are provided (L.P. Finnegan);
f) the long-term use of methadone in heroin addicts has resulted to be safe in terms of side effects, being of no harm on the functions of the liver, and of the endocrine, immune and central nervous systems (M.J. Kreek);
g) M. M. programs facilitate social adjustment, reduce criminality rate (J.J. Déglon, E. Buning, B. Sommer, H. Joseph), decrease engagement in socially disapproved working activities, like prostitution (L. Gunne);
h) finally, M. M. diminishes the risk of contracting infectious diseases (C. Kaplan, M. Reisinger).

Indeed, in a study carried out in New York city from 1983 to 1984, the average incidence of HIV infection in patients on M. M. was less than 10%, at a time when over 50-60% of untreated addicts were infected (M.J. Kreek). In Sweden, admissions for new patients to M. M. programs were closed from 1979 to 1983, due to political controversy. When they were reopened in 1984, the incidence of HIV infection was 3% among old patients, whereas a rapid increase of seropositivity was observed in newly admitted addicts, with a peak of 59% in 1988 (O. Blix).

Such data are of a particular value in view of the fact that the myth held in the 1970's that tuberculosis (TB) could be totally eradicated, vanished with the spreading of HIV epidemics which brought about a new upsurge of TB in the industrialized countries, making this disease the largest cause of death from a single pathogen in the world (J.H. Lowinson).

In conclusion, we have to admit that the alarm consequent to the spread of AIDS epidemics has contributed to the rationalization of the approach to drug addiction more effectively than the most advanced discoveries on the long-term effects of ethanol, cocaine or opioids on neuronal plasticity. However, the firm acquisition that an effective and safe pharmacological therapy specific for opiate addiction does exist, should stimulate the search for medications active on other forms of drug dependence (e.g., alcohol and cocaine)(I. Maremmani).

The diffused idiosyncrasy towards methadone has been partially neutralized by the emotionality induced by the threat of AIDS. However, we should not forget that prejudices are still thriving (A. Coluccia, C. Vetere). In fact, specialists, public health agencies, and laymen still dispute endlessly on whether drug addiction is a disease, and whether only physicians should be responsible for M.M. programs. It must be recognized that such a debate, in light of the reported data on the efficacy of M.M. therapy and of the spreading of HIV epidemics, constitutes, a mere ideological, irrational constraint to correct medical practice and basic research.

GIOVANNI B. CASSANO
Institute of Psychiatry
University of Pisa
Italy

Pisa, November 1994

# CONTENTS

1. **Opposing Tonically Active Endogenous Opioid System Modulating the Mesolimbic Dopamine Reward Pathway.** A. HERZ 1

2. **Recent Advances in the Understanding of Membrane Opioid Receptor.** A. P. SMITH, H. H. LOH 13

3. **Central Dopaminergic Mechanism of Alcohol and Opiate Withdrawal Syndromes.** M. DIANA, Z. L. ROSSETTI, G. L. GESSA 19

4. **Endogenous Morphine and Codeine As Possible Physiological Ligands of μ-Opiate Receptors.** A. TAGLIAMONTE, M. GUARNA, E. BIANCHI 27

5. **Drug of Abuse Craving in Free-Choice Experimental Conditions.** M. SALA, D. BRAIDA, M. COLOMBO, M. P. LEONE, E. GORI 37

6. **Methadone Maintenance. Comes of Age.** V. P. DOLE 45

7. **Pharmacotherapy for craving.** I. MAREMMANI, O. ZOLESI, L. DAINI, M. R. CAPONE, M. AGLIETTI, P. CASTROGIOVANNI 51

8. **The Use of GHB in Anticraving Therapy. Preliminary Experimental Data of its Use in Alcohol and Heroin Dependence.** O. ZOLESI, L. DAINI, M. R. CAPONE, M. AGLIETTI, F. RAIMONDI, I. MAREMMANI. 57

**9. Development of an Italian Version of the "Readiness to Change Questionnaire" for Alcoholism and Addiction** M. SCAGLIA, E. BAIARDO, I. BALDASSO, E. CRISTINA, S. DEODATO, W. O. FAUSTMAN, V. SPILLER, M. STEFANINI, R. TRAMPETTI, G. P. GUELFI 63

**10. Mood Disorders in Heroin Dependence and Clinical Differences Between Heroin Addicts with and without Mood Disorders.** L. DAINI, M. R. CAPONE, T. AGUECI, M. AGLIETTI, O. ZOLESI, I. MAREMMANI 69

**11. Nature and Magnitude of Problems Associated with Psychoactive Drugs.** A. H. GHODSE, S. RAWAF 79

**12. The Street Life and Treatment Response of 105 Heroin Addicted Women.** L. M. GUNNE, L. GRÖNBLADH, L. S. ÖHLUND 91

**13. The Pharmacological Rationale for Methadone Treatment of Narcotic Addiction.** R. G. NEWMAN 109

**14. Reducing Heroin Consumption During Methadone Treatment and Limitation of Post-Treatment Relapses:Two Crucial Public Health Problems.** J. J. DÉGLON 119

**15. Methadone Against AIDS. The Swedish Experience.** O. BLIX 129

**16. Addiction and Pregnancy: Maternal and Child Issues** L. P. FINNEGAN 137

**17. Tuberculosis. The Comeback of a Killer.** J. H. LOWINSON, M. GOUREVITCH 149

**18. Lowering the Threshold: A Comparative Overview of AIDS Prevention Among Injecting Drug Users Using Outreach Technology.** C. D. KAPLAN 157

**19. Pharmacological Treatment of Addiction: Normalization of Physiology and AIDS Risk Reduction.** M. J. KREEK 165

**20. Methadone Treatment and the Epidemiology of AIDS in the European Community.** M. REISINGER 175

**21. Public Health in Face of Drug Addiction and AIDS** C. VETERE 181

**22. The Dark and Light Aspects of Italian Legislation on Substances of Abuse.** A. COLUCCIA 187

**23. Methadone and Criminality. Euro-methwork Workshop.** E. BUNING, P. LANG, M. TORRENS, C. HERZOG, L. GRÖNBLADH, H. JOSEPH, A. TAGLIAMONTE, O. BLIX 193

**24. Workshop of the European Collaborating Centres in Addiction Studies (ECCAS).** 199

**Methadone Substitution Therapy and its Impact on HIV Risk Behaviors.** A.H. GHODSE, C. CLANCY, M. JONES, M. BOURQUIN, B. DE LAS HERAS, M. MONTES, E. KEENAN, I. LOPES, M. RIGLIETTA, L. TIDONE, C. ROSINGER, K. RUNGE NIELSON, L. SAN, F. SCHIFANO, P. SCHOEMANN, J. WERENNE, L. CREMMERS, B. SOMMER,S. WIEVIORKA 200

**A Report from Denmark.** B. SOMMER 203

**A Report from Germany.** C. ROSINGER 204

**A Report from Ireland.** E. KEENAN 205

**A Report from Italy: Methadone Treatment in Opiate Addicts: 12 Years Experience in Italy.** M. RIGLIETTA, R. BAGNATO, M. CAMPANA, P. COLLEONI, P. DRAGO, M. FUSINI, Z. VYHNA'LKOVA, L. TIDONE 206

**The Padova Centre Report.** F. SCHIFANO 207

**A Report from Spain. The Methadone Maintenance Programs in Barcelona.** M. TORRENS, L. SAN 209

**A Report from Switzerland. Methadone Treatment in Switzerland.** M. BOURQUIN 211

**25. 1st European Methadone Conference. Opening Remarks** M.W. PARRINO 213

**26. Incorporating a Methadone Maintenance Program into a Drug-free Treatment System.** M. A. RIZZI 219

**27. Heroin Dependence. Theory of Different Levels of Intervention.** I. MAREMMANI, O. ZOLESI, L. DAINI, R. NARDINI, P. CASTROGIOVANNI 225

## CONTRIBUTORS


| | |
|---|---|
| Aglietti,M. | Institute of Psychiatry, University of Pisa, Italy |
| Agueci,T. | Institute of Psychiatry, University of Pisa, Italy |
| Bagnato,R. | SERT, USSL 29, Bergamo, Italy |
| Baiardo,E. | SERT, USL 3 "Genovese", Genova, Italy |
| Baldasso,I. | SERT, USL 3 "Genovese", Genova, Italy |
| Bianchi,E. | Institute of Pharmacology, University of Siena, Italy |
| Blix,O. | Department of Alcohol and Drug Diseases, Malmö General Hospital, Sweden |
| Bourquin,M. | The Phenix Foundation, Geneva, Suisse |
| Braida,D. | Institute of Pharmacology, Faculty of Mathematical, Physical and Natural Sciences, University of Milan, Italy |
| Buning,E. | Euro-methwork, GC&GD, 1018 WT Amsterdam, Holland |
| Campana,M. | SERT, USSL 29, Bergamo, Italy |
| Capone,M.R. | Institute of Psychiatry, University of Pisa, Italy |
| Castrogiovanni,P. | Institute of Psychiatry, University of Pisa, Italy |
| Clancy,C. | ECCAS, London, U.K. |
| Colleoni,P. | SERT, USSL 29, Bergamo, Italy |
| Colombo,M. | Institute of Pharmacology, Faculty of Mathematical, Physical and Natural Sciences, University of Milan, Italy |
| Coluccia,A. | Departement of Legal Medicine, University of Siena, Italy |
| Cremmers,L. | ECCAS, Brussels, Belgium |
| Cristina,E. | SERT, USL 3 "Genovese", Genova, Italy |
| Daini,L. | Ph.D. Research Program, Siena,Cagliari & Pisa Universities, Italy |
| De las Heras,B. | ECCAS, Oviedo, Spain |
| Déglon,J.J. | The Phenix Foundation, Geneva, Suisse |
| Deodato,S. | SERT, USL 3 "Genovese", Genova, Italy |

| | |
|---|---|
| Diana,M. | "Bernard Brodie" Department of Neurosciences, University of Cagliari, Italy |
| Dole,V.P. | The Rockfeller University, New York, NY, U.S.A. |
| Drago,P. | SERT, USSL 29, Bergamo, Italy |
| Faustman,W.O. | VA Medical Center and Stanford University School of Medicine, Palo Alto, CA, U.S.A. |
| Finnegan,L.P. | Women's Health Initiative, National Institute of Health, Bethesda, MD, U.S.A. |
| Fusini,M. | SERT, USSL 29, Bergamo, Italy |
| Gessa,G.L. | "Bernard Brodie" Department of Neurosciences, University of Cagliari, Italy |
| Ghodse,A.H. | Division of Addictive Behaviour, St George's Hospital Medical School, London, U.K. |
| Gori,E. | Institute of Pharmacology, Faculty of Mathematical, Physical and Natural Sciences, University of Milan, Italy |
| Gourevitch,M. | Albert Einstein College of Medicine of Yeshiva University, Division of Substance Abuse, Bronx, New York, NY, U.S.A. |
| Grönbladh,L. | Departement of Psychiatry, University of Uppsala, Sweden |
| Guarna,M. | Institute of Histology and General Embryology, University of Siena, Italy |
| Guelfi.G.P. | SERT, USL 3 "Genovese", Genova, Italy |
| Gunne,L.M. | Departement of Psychiatry, University of Uppsala, Sweden |
| Herz,A. | Department of Neuropharmacology, Max-Planck-Institute for Psychiatry, Martinsried, Germany |
| Herzog,C. | Euro-methwork, GC&GD, 1018 WT Amsterdam, Holland |
| Jones,M. | ECCAS, London, U.K. |
| Joseph,H. | The Chemical Dependency Research Working Group, New York, NY, U.S.A. |
| Kaplan,C.D. | Institute for Psychosocial and Socioecological Research, University of Limburg, Maastricht, The Netherlands |
| Keenan,E. | Drug Treatment Centre, Dublin, Ireland |
| Kreek,M.J. | Thr Rockerfeller University, New York, NY, U.S.A. |
| Lang,P. | Euro-methwork, Bremen, Germany |
| Leone,M.P. | Institute of Pharmacology, Faculty of Mathematical, Physical and Natural Sciences, University of Milan, Ital |
| Loh,H.H. | Department of Pharmacology, University of Minnesota Medical School, Minneapolis, MN, U.S.A. |
| Lopes,I. | ECCAS, Porto, Portugal |
| Lowinson,J.H. | Albert Einstein College of Medicine of Yeshiva University, Division of Substance Abuse, Bronx, New York, NY, U.S.A. |

| | |
|---|---|
| Maremmani,I. | Dependence-Psychiatry Unit, Institute of Psychiatry, University of Pisa, Italy |
| Montes,M. | ECCAS, Oviedo, Spain |
| Nardini,R. | Social Disease Study and Intevention Association, Pietrasanta, Italy |
| Newman,R.G. | Beth Israel Medical Center, New York City, NY, U.S.A. |
| Öhlund,L.S. | Department of Psychiatry, University of Uppsala, Sweden |
| Parrino,M.W. | American Methadone Treatment Association, New York City, NY, U.S.A. |
| Raimondi,F. | Institute of Psychiatry, University of Pisa, Italy |
| Rawaf,S. | Division of Addictive Behaviour, St George's Hospital Medical School, London, U.K. |
| Reisinger,M. | Commission of the European Community, European Monitoring Centre for Drugs and Drug Addiction, (EMCDDA), Brussels, Belgium |
| Riglietta,M. | SERT, USSL 29, Bergamo, Italy |
| Rizzi,M.A. | CODAC, Cranston, Rhode Island, U.S.A. |
| Rösinger,C. | Rheinische Landes-und Hochschulklinik, Essen, Germany |
| Rossetti,Z.L. | "Bernard Brodie" Department of Neurosciences, University of Cagliari, Italy |
| Runge Nielson,K. | ECCAS, Arhus, Denmark |
| Sala,M. | Institute of Pharmacology, Faculty of Mathematical, Physical and Natural Sciences, University of Milan, Italy |
| San,L. | CAS-Barceloneta, University of Barcelona, Spain |
| Scaglia,M. | SERT, USL 3 "Genovese", Genova, Italy |
| Schifano,F. | SERT, ULSS 21, Padova, Italy |
| Schoemann,P. | ECCAS, Brussels, Belgium |
| Smith,A.P. | Department of Pharmacology, University of Minnesota Medical School, Minneapolis, MN, U.S.A. |
| Sommer,B. | Embedslaege Institutionen, For Ringkobing, AMT, Ringkobing, Denmark |
| Spiller,V. | SERT, USL 3 "Genovese", Genova, Italy |
| Stefanini,M. | SERT, USL 3 "Genovese", Genova, Italy |
| Tagliamonte,A. | Institute of Pharmacology, University of Siena, Italy |
| Tidone,L. | SERT, USSL 29, Bergamo, Italy |
| Torrens,M. | CAS-Barceloneta, University of Barcelona, Spain |
| Trampetti,R. | SERT, USL 3 "Genovese", Genova, Italy |
| Vetere,C. | UNICRI, Rome, Italy |
| Vyhnalkova,Z. | SERT, USSL 29, Bergamo, Italy |

Werenne, J. — SERT, USSL 29, Bergamo, Italy

Wieviorka, S. — Centre Saint-Germain-Pierre-Nicole, Croix-Rouge Française, Paris, France

Zolesi, O. — Ph.D. Research Program, Siena, Cagliari & Pisa Universities, Italy

**Acknowledgment**

We wish to express our gratitude to

Associazione per l' Utilizzo delle Conoscenze Neuroscientifiche a fini Sociali

AU-CNS

*Istituto di Farmacologia, Università di Siena*
*Via delle Scotte, 6 53100 Siena*
*Via XX Settembre, 83 55045 Pietrasanta (LU)*
*S. Margherita la Suvera, 7 53010 Vagliagli (SI)*
*Italy*

for editing this book.

# 1

# Opposing Tonically Active Endogenous Opioid System Modulating the Mesolimbic Dopamine Reward Pathway

**A. Herz**
Department of Neuropharmacology, Max-Planck-Institute for Psychiatry, Martinsried, Germany

## Summary

Opioids, prototypical drugs of abuse, exert marked effects on motivation. μ- and ∂-opioid receptor agonists function as positive reinforcers. In contrast, k-opioid receptor agonists induce aversive and dysphoric states. The conditioned place preference paradigm was used to examine the neuroanatomical substrates mediating the motivational effects of highly selective opioids and the technique of in-vivo microdialysis was applied to assess the neurochemical basis of such effects. Conditioned place preference was obtained by microinjection of the μ-opioid receptor agonist DAGO into the ventral tegmental area (VTA), (the site of origin of the A10 dopamine neurons), but not when injected into the nucleus accumbens (NAC) and in further forebrain areas. In contrast, conditioned place aversion was obtained by microinjection of the specific k-opioid receptor agonists U-50,488 and/or U-69593 into the VTA, NAC, medial frontal cortex and lateral hypothalamus; microinjection of these μ- and k-agonists into the substantia nigra and the caudatus putamen were ineffective. Microinjection of DAGO into the VTA resulted in significant increase in the release of dopamine (DA) and its metabolism in the NAC. However, DAGO infusions into the NAC via the microdialysis probe did not affect release of DA or its metabolites. The selective blockade of μ-receptors by CTOP within the VTA produced significant decrease of DA basal release. In contrast, CTOP infusions into the NAC were without effect. The selective k-agonist U 69853 microinjected into the VTA did not change the DA overflow. In contrast, U 69853 infusion into the NAC resulted in a dose-dependent significant decrease in DA release. Infusions of the k-antagonist nor-BNI into the NAC dose-dependently increased basal DA release within the NAC, but it was ineffective when injected into the VTA. These data demonstrate that the same mesolimbic pathway is modulated by two opposing tonically active endogenous opioid systems. The concerted action of the two systems (possibly a ß-endorphin system in the VTA and a dynorphin presynaptically located dynorphin system in the NAC) allows the maintenance of basal mesolimbic DA release.

## Introduction

Opioids and other drugs of abuse exert marked effects on mood and motivation (1,2). They produce euphoria in humans and are self-administered in animals.

Repeated administration of opioids results in the development of tolerance and physical dependence.

Although these effects of chronic drug intake may be important for the maintenance of drug addiction once established, it is now generally accepted that they do not represent causal factors for the development of addiction. There is now ample evidence that the positive reinforcing effects caused by activation of endogenous reward pathways determine the abuse potential of certain drugs and initiate the addictive process. It is of particular interest that the same neural systems that mediate the reinforcing or motivational effects of opioids may be shared by other drugs of abuse; these neural pathways are also involved in natural reward processes (3-5).

## Bidirectional Motivational Effects of Opioids

In the behavioural experiments discussed here, the drug-induced motivational effects were assessed by place conditioning, a technique which measures the secondary reinforcing effects of drugs. In this procedure, the association that develops between the presentation of a drug and a previously neutral stimulus (e.g. differently colored compartments of a shuttle box) is evaluated. The results obtained with this paradigm for rewarding drugs are largely identical with those obtained with the self-administration paradigm technique (6). Besides the evaluation of rewarding properties, the place conditioning procedure also allows the detection of aversive (negatively reinforcing) properties of drugs, which is of particular importance when investigating the effects of opioids.

Depending on the receptor specificity of the opioid, rewarding or aversive effects were demonstrated: dose-dependent place preference induced by systemic application of μ-opioid receptor ligands as morphine, fentanyl or sufentanyl; place aversion by systemic application of the k-receptor specific benzenacetamide-derivatives U 50,488 and U-69593 (7) (Fig. 1).

Similarly, rewarding or aversive effects of μ- and k-receptor ligands were also obtained by intracerebroventricular (i.c.v.) injection of much lower doses, indicating the central origin of these motivational effects (8). I.c.v. application also allowed testing of the effects of (hydrophilic) opioid peptides, having little access to brain: DAGO; a highly specific μ-receptor ligand induced place preference, while E-2078, a metabolically stable dynorphin-like peptide (9), induce aversion. DPDPE, a ∂-selective enkephalin derivative caused dose-dependent place preference, indicating that besides μ-receptors, ∂-receptors also mediate rewarding effects (10).

ß-endorphin (B-EP) is of particular interest as it represents the main opioid peptide derived from pro-opiomelanocortin (POMC). I.c.v. application of ß-EP as a conditioning drug induced dose-dependent place preference (11,12). This effect was antagonized by the specific μ-receptor antagonist CTOP as well as by the specific ∂-receptor antagonist ICI 164,864. These results are in line with the results of in vitro experiments,

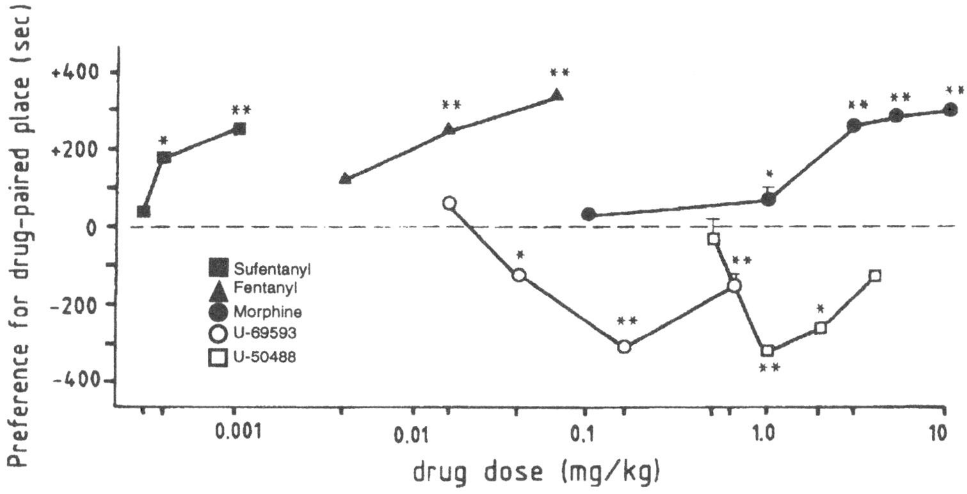


**Fig. 1.** Place conditioning produced by the μ-opioid receptor agonists morphine, fentanyl and sufentanyl and the K-opioid receptor agonists U50,488 and U-69593. Ordinate: mean difference (seconds) between time spent in drug-and vehicle-paired sides of the test box. Abscissa: drug dose. Points above zero indicate place preference, points below place aversion. Each point represents the mean conditioning score of eight to ten rats. Asterisks denote significant place conditioning (Wilcoxon test * p<.05; ** p<.01). Adapted from (7)

showing that ß-EP exhibits high affinity binding to μ- as well as ∂-opioid receptors (13).

## Endogenous Opioidergic Tone

Of particular significance in view of the existence of an endogenous opioidergic tone in the reward system are experiments in which opioid antagonists were used as conditioning drugs. Systemic as well as i.c.v. injection of the universal opioid antagonists naloxone or naltrexone induced strong place aversion (7). A similar effect was obtained with the μ-receptor specific antagonist CTOP, while the ∂-specific antagonist ICI 174,864 lacked such an effect. Treatment with the k-specific antagonist nor-NBI resulted in a tendency towards place preference which, however, did not always prove significant (8). Stereospecific aversive properties of narcotic antagonists preferring the μ-receptor have been described in rats and monkeys by others (14-16). Thus, there is good evidence for the existence of a tonically active reward system mediated by μ-receptors. The finding that the enkephalins inhibitor thiorphan induced rewarding effects in such experiments indicates that activation of ∂-receptors (by enkephalins) may also contribute to an endogenous reward tone (17).

A likely candidate as endogenous ligand of the reward pathway mediated by μ- and ∂-receptors is ß-EP. The particular anatomy of the cerebral ß-EP system - the great majority of ß-EP fibers originate from the nucleus arcuatus of the hypothalamus - makes

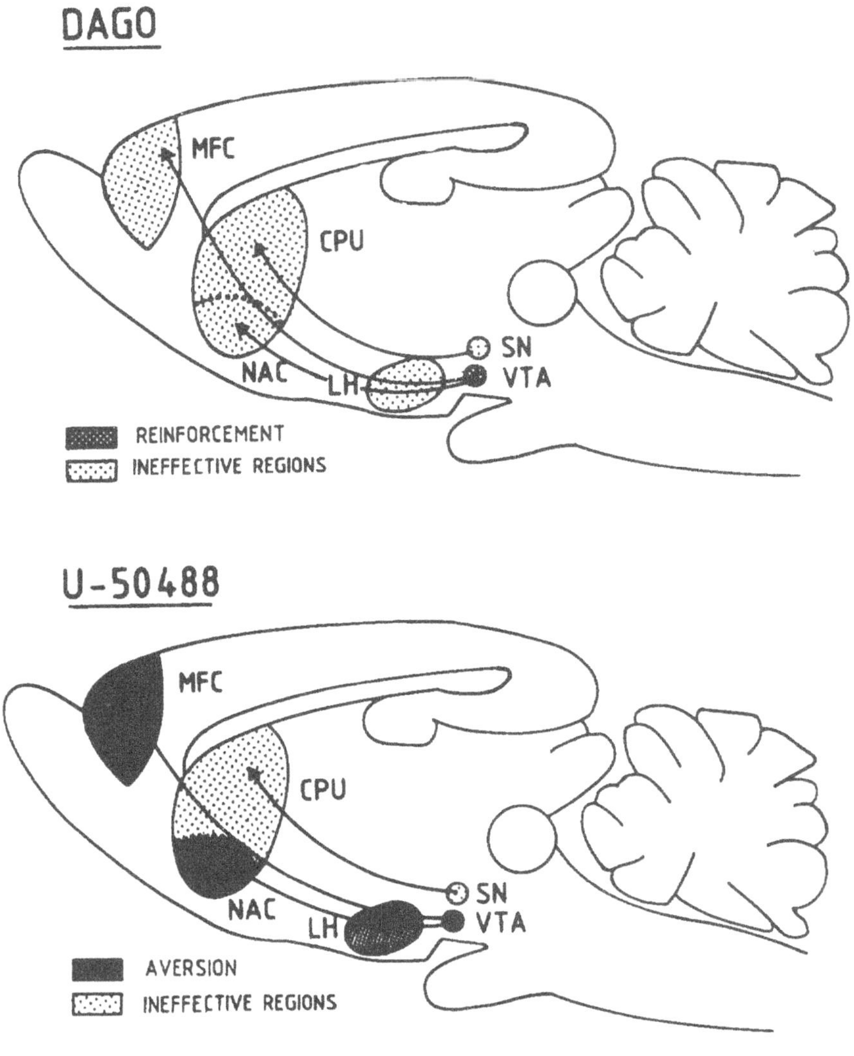


**Fig. 2.** Brain regions implicated in the mediation of the reinforcing and aversive effects of opioids as revealed by microinjection of the μ-receptor ligand DAGO (above) and the k-receptor ligand U 50488 (below) in the rat brain. Microinjection of DAGO induced place preference only in the ventral tegmentum area (VTA), microinjection of U 50488 induced place aversion in the VTA, the nucleus accumbens (NAC), the medial frontal cortex (MFC) and in the lateral hypothalamus (LH). Both substances were without effect in the substantia nigra (SN) and the caudatus putamen (CPU). Adapted from (7)

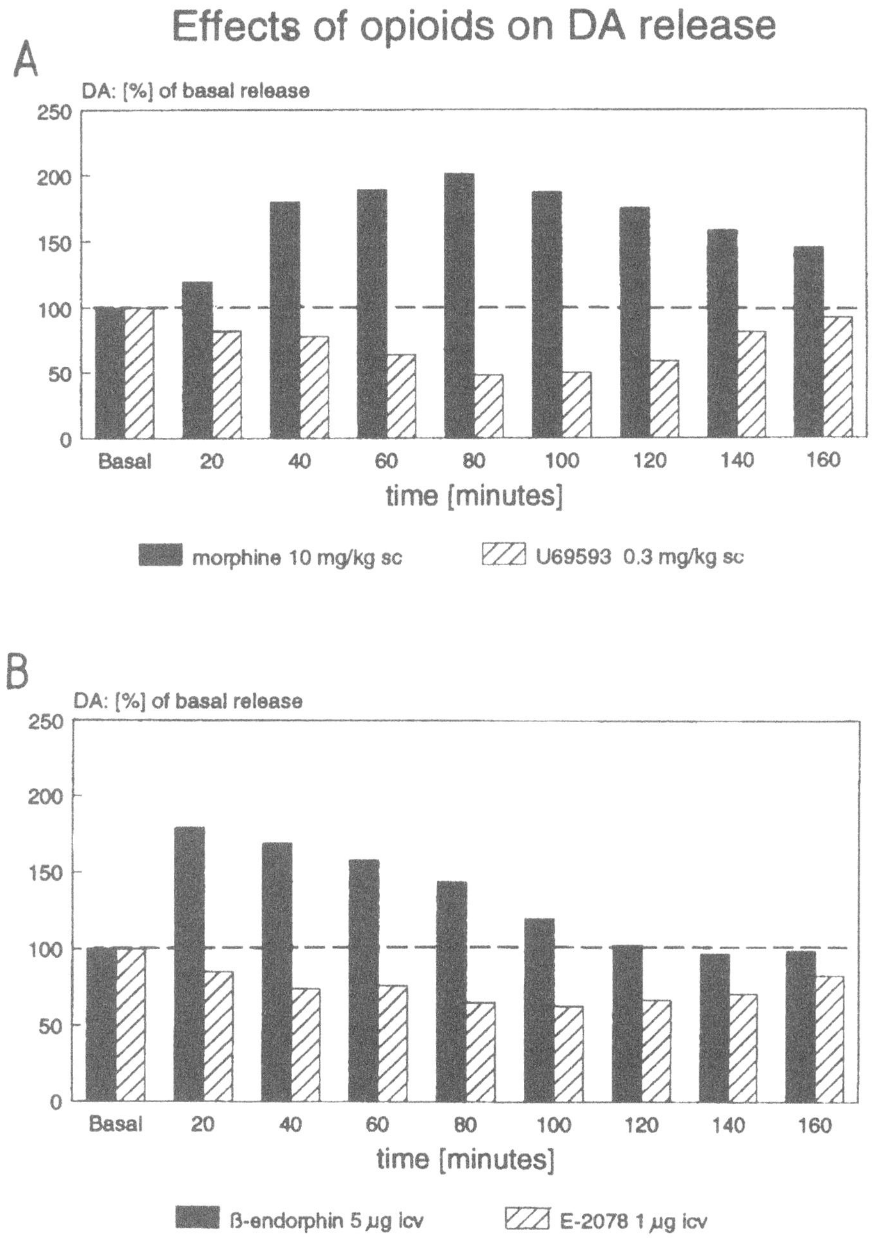


**Fig. 3.** Effects of pheriferal (above) or central (below) administered µ-opioid agonists (morphine, ß-endorphin) and k-opioid agonists (U-69593, dynorphin analog E-2078) on DA release in the NAC. Data are expressed as the mean percentages of four dialysates (basal). Adapted from (30)

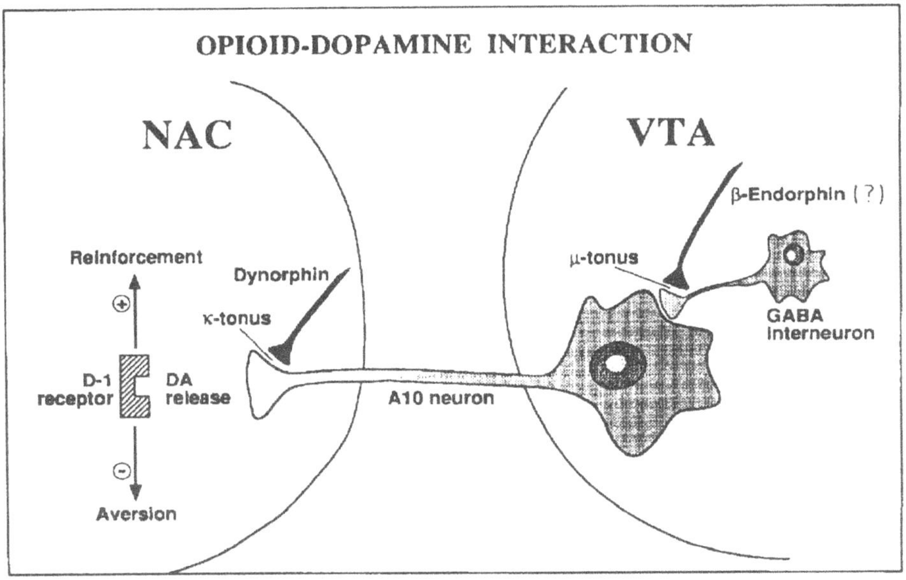


**Fig. 4.** Model for the modulation of mesolimbic A10 neurons by endogenous opioid systems. The tonic stimulation of μ-receptors in the VTA by a ß-endorphin system and of k-receptors in the NAC by a dynorphin system is necessary to maintain basal DA release. Adapted from (33)

it possible to largely eliminate this system by lesioning of this hypothalamic nucleus. In place conditioning experiments in arcuate-lesioned rats the aversive effects of naloxone were found to be largely abolished, while the rewarding effects of morphine and the aversive effects of U 50,488 remained unchanged. This strongly supports the notion that a continuous reward tone is caused by presynaptic release of ß-EP from fibers arising from the hypothalamus (18). This is in line with other experiments from which an important role of this opioid peptide in reward processes was deduced (2,19-21).

## D1 Dopamine Receptors in Reward Processes

Findings showing that motivational effects of opioids and other drugs of abuse are affected by manipulation of dopamine (DA) neurotransmission suggest that DA neurons are key components of reward pathways (22,23). The identification of distinct DA receptor subtypes and the availability of selective antagonists for the D 1 and D 2 receptors offered the possibility to reevaluate the DA hypothesis of reward and to investigate the involvement of particular receptor subtypes in the mediation of reinforcing and aversive states. Acute systemic administration of selective D 1 DA antagonist SCH 23390 caused clear conditioned place aversion, whereas the specific D 2-antagonists sulpiride and spiperone lacked such effects. These findings support the hypothesis that there is a tonic activation of D 1 DA-receptor-mediated pathways, the disruption of which results in aversive states (24-26).

## Brain Sites Involved in Reward and Aversion

The use of the microinjection technique permitted the identification of the specific brain areas involved in the reinforcing properties of opioids. Both, self-administration and place conditioning studies, revealed the injection of morphine to enkephalin analogues into several brain areas, including the hypothalamus, septum, ventral tegmental area (VTA) and nucleus accumbens (NAC) to be rewarding (27). The VTA and NAC represent key structures of the mesolimbic DA system. The DA perikarya area located in the VTA, while the NAC represents an important terminal field of the fibers arising therein. The doses required to produce reward were lowest in the VTA, suggesting that this structure may be of critical importance in this respect (23). In contrast to reward, much less was known, until recently, on brain areas involved in the aversive effects of opioids.

In recent microinjection studies, the neuroanatomical sites of action of μ- and k-opioid receptor agonists were compared.

Microinjection of DAGO induced conditioned place preference only in the VTA, but not in the NAC, medial frontal cortex (MFC) or lateral hypothalamus (LH). In contrast, microinjection of U 50,488 and E-2078 induced place aversion in the VTA, NAC, MFC and LH. Both, μ- and k-receptors agonists were ineffective when injected into the substantia nigra (SN) and the caudatus-putamen (CPU) (Fig. 2).

These data suggest that rewarding effects are associated with the activation of μ-receptors in the VTA, whereas the aversive effects may originate not only from the VTA, but also from limbic-cortical terminal fields of the reward system (28). Further microinjection studies addressed the question of whether the NAC-mediated rewarding effects of μ-receptor ligands, and the aversive effects of k-receptor ligands, necessarily involve dopaminergic mechanisms. Inactivation of these by local injection of the neurotoxin 6-OHDA or the D-1-DA antagonist SCH 23390 into the NAC abolished the rewarding as well as the aversive effects of μ- and k-receptor ligands, respectively.

These data indicate that DA neural transmission in the NAC is necessary not only for the manifestation of the rewarding, but also of the aversive effects of opioids (29).

## Microdialysis Studies

This in-vivo method allows measurement of the release of minimal amounts of neurotransmitter substances from circumscribed brain areas. In the present experiments, microdialysis probes were implanted in the NAC, and the release of DA and its main metabolites HVA and DOPAC was measured by means of high performance liquid chromatography.

A dose-dependent increase in the release of DA, HVA and DOPAC was obtained by systemic injection of morphine and i.c.v. injection of the μ-receptor agonist DAGO, the ∂-receptor agonist DPDPE as well as by ß-EP; the k-receptor ligand U 69593 induced decrease of release in the NAC upon systemic and i.c.v. injection. The dynorphin analogue E-2078 had a similar effect upon i.c.v. administration (Fig. 3). The effective doses in these experiments were very similar to those in the above behavioural tests (30-32).

Most interesting results on the modulation of the mesolimbic DA system were obtained when the opioids were either microinjected into the VTA (via chronically implanted cannula) or infused into the NAC (through the microprobe from which the samples for evaluation of DA and its metabolites were also taken). Microinjection of the μ-receptor agonist DAGO into the VTA increased DA release in the NAC, while CTOP decreased basal DA levels. The k-receptor agonist U 69593 and the k-receptor antagonist nor-BNI were ineffective at this site. Quite opposite results were obtained when these drugs were applied (by infusion through the microprobe) into the NAC: DAGO and its antagonist CTOP did not change basal DA release, while U 69593 decreased, and nor-BNI increased, basal DA levels. These data indicate basal DA release to be under the tonic control on both opioid systems: μ-receptor activation in the VTA results in increase of DA release, whereas k-receptor activation in the NAC decreases basal DA release (33).

A synopsis of these results is given in a model which illustrates the modulation of the mesolimbic DA neurons ascending from the VTA to the NAC (Fig. 4).

There are several reports showing that local application of morphine into the VTA increases discharge activity in the mesolimbic pathways (34). Such an excitatory effect of opioids (which are basically inhibitory) are most easily explained by an interaction with inhibitory GABA-ergic interneurons, whose existence in the VTA is well documented (35). Lesion experiments of the nucleus arcuatus showed, as discussed above, that the ß-EP reward pathways is tonically active. Although not directly proven, it is tempting to speculate that ß-EP terminals make presynaptic contact with GABA interneurons in the VTA. In any case, behavioural as well as neurochemical data indicate that the mesolimbic DA pathway, ascending from the VTA, is tonically activated by opioid peptides interacting with μ- and possibly ∂-receptors.

In the NAC, the inhibitory effect of ∂-receptor agonists in convincingly explained by ∂-receptors located on the terminals of the mesolimbic DA neurons. In vitro data support such a location and action (36). These receptors seem also to be tonically activated, as judged from the increase in DA release following treatment with the k-receptor antagonist nor-BNI. This suggests that the maintenance of basal DA release in the NAC depends on a balance between the stimulatory μ- (probably also -) opioid system and the inhibitory k-opioid system.

## Concluding Remarks

The data presented here implicate that the motivational effects of opioids are closely linked with the mesolimbic DA system. This became particularly obvious from experiments which showed that the rewarding as well as aversive effects of opioids were dependent on intact DA-ergic neurotransmission in the NAC (29). There are, however, some data in the literature which tend to suggest that DA pathways are not absolutely necessary for the expression of the rewarding effects of opioids (37,38). Such obvious contradictions may possibly be explained by the distinction of "primary incentive" and "consummatory" properties of opioids (5). According to this concept the incentive, but not the consummatory reward mechanisms require intact DA-ergic transmission. The place conditioning paradigm as used in the present experiments may be expected to reveal the incentive properties of a drug, whereas self-administration for instance, may disclose its consummatory properties.

Interesting support for the mediation of opioid reward by the mesolimbic DA system comes from experiments, in which rats were withdrawn from the drug after chronic morphine treatment. They revealed a significant depression of DA release in the NAC upon spontaneous withdrawal (5,39,40). This finding is in line with the general view that withdrawal signs represent the mirror-image of acute opioid effects. This withdrawal-induced decrease in DA release matches with the acute effects of k-receptor ligands which are known to cause dysphoria and anxiety (41), phenomena which are characteristic for morphine withdrawal also. Of particular interest is the finding that this same mesolimbic pathway is modulated by two opposing tonically active endogenous opioid systems. The concerted action of the two systems (a ß-EP system in the VTA, and a dynorphin system in the NAC) allows the maintenance of basal mesolimbic DA release. It is suggested that normal basal activity of the mesolimbic reward pathway is prerequisite for an equilibrium state of emotion and mood.

**References**


1. Stein L, Belluzzi J D. Brain endorphins: possible mediation of pleasurable states. In: Endorphins in Mental Health Research, edited by Usdin E, Bunney W E, Kline N S. McMillan, London, 1977, pp 375-389.
2. Dum J, Herz A. Opioids and motivation. Interdisciplinary Sci Rev 12:180-189, 1987.
3. Kornetsky C, Esposito R U. Euphorigenic drugs: Effects on the reward pathways of the brain. Federation Proc 38:2473-2476, 1979.
4. Herz A, Shippenberg T S. Motivational effects of opioids: neurochemical and neuroanatomical substrates. In: Molecular and cellular aspects of the drug addictions, edited by Goldstein A. Springer, Heidelberg, 1989, pp 111-141.
5. DiChiara G, North R A. Neurobiology of drug abuse. TIPS 13:185-191, 1992.
6. Carr G D, Fibinger H C, Phillips A G. Conditioned place preference as a measure of drug reward. In: The neuropharmacological basis of reward, edited by Liebman J M, Cooper S J. Sci. Publ. Press, Oxford, 1989, pp 264-319.
7. Mucha R F, Herz A. Motivational properties of K-and μ-opioid agonists studies with place and taste preference conditioning procedure. Psychopharmacology 86:274-280, 1985.
8. Bals-Kubik R, Herz A, Shippenberg T S. Evidence that the aversive effects of opioid antagonists and K-agonists are centrally mediated. Psychopharmacology 98:203-206, 1989.
9. Tachibana S, Oshino H, Arakawa Y, Nakzawa T, Araki S, Kaneko T, Yamatsu K, Miyagawa H. Design and synthesis of metabolically stable analogs of dynorphin-A and their analgesic characteristics. In: Biowarning System in the Brain, edited by Takagi H, Oomura Y, Ito M, Otsuka M. University of Tokyo Press, Tokyo, 1988, pp 101-110.
10. Shippenberg T S, Bals-Kubik R, Herz A. Motivational properties of opioids: Evidence that an activation of ∂-receptors mediates reinforcement processes. Brain Res 436:234-239, 1987.
11. Amalric M, Cline E J, Martinez J L, Bloom F E, Koob G F. Rewarding properties of ß-endorphin as measured by conditioned place preference. Psychopharmacology 91:14, 1987.
12. Bals-Kubik R, Shippenberg T S, Herz A. Involvement of μ- and ∂-receptors in mediating the reinforcing effects of ß-endorphin in the rat. Psychopharmacology 175:63-69, 1990.

13. Magnan J, Paterson S J, Tavani A, Kosterlitz H W. The binding spectrum of narcotic analgesic drugs with different agonist and antagonist properties. Naunyn Schmiedeberg's Arch Pharmacol 319:197-205, 1982.
14. Stolerman I P, Pilcher C W T, D'Mello G D. Stereospecific aversive properties of narcotic antagonists in morphine-free rats. Life Sci 22:1755-1762, 1978.
15. Downs D A, Woods J H. Naloxone as a negative reinforcer in rhesus monkeys: effects of dose, schedule, and narcotic regimen. Pharmacol Rev 27:397-405, 1976.
16. Dymshitz J, Lieblich I. Opiate reinforcement and naloxone aversion, as revealed by place preference paradigm, in two strains of rats. Psychopharmacology 92:473-477, 1987.
17. Glimcher P W, Giovino A A, Margolin D H, Hoebel B G. Endogenous opiate reward induced by an enkephalin inhibitor, thiorphan injected into the ventral midbrain. Behavioral Neuroscience 98:262-268, 1984.
18. Mucha R F, Millan M J, Herz A. Aversive properties of naloxone in non-dependent (naive) rats may involve blockade of central ß-endorphin. Psychopharmacology 86:281-285, 1985.
19. Van Ree J M, Smyth D G, Colpaert F C. Dependence creating properties of lipotropin C-fragment (ß-endorphin): evidence for its internal control of behavior. Life Sci 24:495-502, 1979.
20. Dum J, Gramsch C, Herz A. Activation of hypothalamic ß-endorphin pool by reward induced by highly palatable food. Pharmacological Biochemical Behavior 18:443-447, 1983.
21. Sweep C G J, Wiegant V M, DeVry J, Van Ree J M. ß-endorphin in brain limbic structures as neurochemical correlates of psychic dependence on drugs. Life Sci 44:1133-1140, 1989.
22. Bozarth M A, Wise R. Heroin reward is dependent on a dopaminergic substrate. Life Sci 29:1881-1886, 1981.
23. Wise R A, Rompre P P. Brain dopamine and reward. Ann Rev Psychol 40:191-225, 1989.
24. Shippenberg T S, Herz A. Motivational effects of opioids: Influence of D-1 versus D-2 receptor antagonists. Eur J Pharmacol 103:209-214, 1991.
25. Shippenberg T S, Bals-Kubik R, Huber A, Herz A. Neuroanatomical substrates mediating the aversive effects of D1 dopamine receptor antagonists. Psychopharmacology 103:209-214, 1991.
26. Acquas E, Carboni E, Leone P, DiChiara G. SCH 23390 blocks drug conditioned place-preference and place aversion: anhedonia (lack of reward) or apathy (lack of motivation) after dopamine receptor blockade. Psychopharmacology 99:151-155, 1989.
27. Wise R A, Bozarth M A. Action of drugs of abuse on brain reward systems: an update with specific attention to opiates. Pharmacol Biochem Behav 17:239-243, 1982.
28. Bals-Kubik R, Ableitner A, Herz A, Shippenberg T S. Neuroanatomical sites mediating the motivational effects of opioids as mapped by the conditioned place preference paradigm in rats. J Pharmacol Exp Ther 264:489-495, 1993.
29. Shippenberg T S, Bals-Kubik R, Herz A. Examination of the neurochemical substrates mediating the motivational effects of opioids: role of the mesolimbic dopamine system and D-1 vs. D-2 dopamine receptors. J Pharmacol Exp Ther 265:53-59, 1993.
30. Spanagel R, Herz A, Shippenberg T S. The effects of opioids peptides on dopamine releasr in the nucleus accumbens: an in-vivo microdialysis study. J Neurochem 55:1734-1740, 1990
31. Spanagel R, Herz A, Shippenberg T S. Identification of the opioid receptor types mediatin ß-endorphin-induced alterations in dopamine release in the nucleus accumbens. Eur . Pharmacol 190:177-184, 1990.
32. DiChiara G, Imperato A. Opposite effects of μ- and - opioid agonists on dopamine releas in the nucleus accumbens and dorsal caudate of freely moving rats. J Pharmacol Exp The 244:1067-1080, 1988.
33. Spanagel R, Herz A, Shippenberg T S. Opposing tonically active endogenous opioid system modulate dopaminergic pathways. Proc Natl Acad Sci 89:2040-2050, 1992.
34. Gyslling K, Wang R Y. Morphine-induced activation of A10 dopamine neurons in the ra

Brain Res 277:119-127, 1983.
35. Kalivas P W, Duffy P, Eberhardt H. Modulation of A10 dopamine neurons by gamma-aminobutyric acid agonists. J Pharmacol Exp Ther 253:858-866, 1990.
36. Heijina M H, Padt M, Hagenboom F, Portoghese P S, Mulder A H, Schoffelmeer A N M. Opioid receptor mediated inhibition of dopamine and acetylcholine release from slices of rat nucleus accumbens, olfatory tubercle and frontal cortex. Eur J Pharmacol 18:267-278, 1990.
37. Ettenberg A, Pettit H G, Bloom F E, Koob G F. Heroin and cocaine intravenous self-administration in rats: mediation by separate neural systems. Psychopharmacology 78:204-209, 1982.
38. Stinus L, Cador M, Le Moal M. Interaction between endogenous opioids and dopamine within the nucleus accumbens. In: The Neurobiology of Drug and Alcohol Addiction, edited by Kalivas P W, Samson N H. Ann. N.Y. Academy of Science, New York, 1992, pp 254-273.
39. Acquas E, Carboni E, Leone P, DiChiara G. Profound depression of mesolimbic dopamine release after morphine in dependent rats. Eur J Pharmacol 193:133-134, 1991.
40. Rossetti L L, Hamaidan Y, Gessa G L. Marked inhibition of mesolimbic dopamine release: a common feature of ethanol, cocaine and amphetamine abstinence in rats. Eur J Pharmacol 22&:226-234, 1992.
41. Pfeiffer A, Brantl V, Herz A, Emrich H G. Psychotomimesis mediated by K-opiate receptors. Science 233:774-776, 1986.

# 2

# Recent Advances in the Understanding of Membrane Opioid Receptor

**A. P. Smith, H. H. Loh**
Department of Pharmacology, University of Minnesota Medical School Minneapolis, MN, USA

**Summary**

The cloning of the ∂ opioid receptor by two independent groups has been quickly followed by cloning of the other two major opioid receptor types, μ and k. It is likely that other opioid receptor types of subtypes will soon be discovered, and the availability of the receptor cDNAs will also allow investigators to explore their localization and function. Our own laboratory, following a more traditional approach, has isolated a protein that seems to play some role in opioid receptor function.

Opioid receptors mediate the effects of exogenous, analgesic alkaloids such as morphine, as well as the endogenous opioid peptides, which include ß-endorphin, enkephalins and dynorphin. They were one of the first class of cell surface receptors to be identified using radioligand binding assays (1-3), yet proved to be extraordinarily difficult to purify. This was primarily because of difficulties in solubilizing them in a form retaining ligand binding activity, but also because of their heterogeneity, and the lack of a simple biochemical test that can demonstrate the functional relevance of isolated receptors (4).

The very recent reports of successful cloning of an opioid receptor (5,6) have finally brought this field to the point where the highly sophisticated tools of molecular biology can be used to explore opioid receptor structure and function, as has been done with other cell surface receptors. In this article, we briefly review this work and its implications, as well as its relationship to our own studies of opioid receptors.

## Cloning of Opioid Receptors

At the end of last year, two laboratories simultaneously reported cloning of the ∂ opioid receptor from NG108-15 neuroblastoma x glioma cells (5,6), which are believed to contain exclusively this type. The two laboratories used a similar approach, preparing a cDNA library from NG108-15 cells, transfecting pools of this cDNA into COS cells, then assaying the cells for binding of a radioactive opioid ligand. The cloned, expressed receptor showed typical binding properties expected of an opioid receptor, including high affinity, stereoselectivity, and preference for a particular class of opioids, namely ∂; in addition, Evans, et al. (5) demonstrated that binding to the expressed receptor inhibited adenyl cyclase, as is the case with native ∂ opioid receptors on NG108-15 cells.

In NG108-15 cells (7) as well as in at least some areas of the mammalian CNS (8-11) opioid receptors are coupled to G-proteins, so it was no surprise that the predicted amino acid sequence of the cloned opioid receptor cDNA was homologous to other members of the G-protein coupled receptor superfamily (12) in particular, the sequence displayed seven characteristic hydrophobic regions which are presumed to span the cell membrane. The cloned ∂ opioid receptor cDNA possesses several other features in common with other G-protein coupled receptors, including a) aspartate residues in characteristic positions in transmembrane regions two and three which are thought to be involved in ligand binding and coupling to adenyl cyclase; b) cysteine residues in extracellular loops two and three, which may form a disulfide bridge between these regions; c) several putative glycosylation sites in the N-terminal region; and d) several putative phosphorylation sites in the C-terminal region, which in other G-protein coupled receptors are thought to be involved in regulation of the receptor during chronic agonist treatment. Among the other members of this superfamily, the somatostatin receptors are most closely related to the ∂ opioid receptor, exhibiting 35-40% sequence homology, and this may be related to the report that somatostatin ligands can bind to opioid receptors.

Whenever the cDNA for a receptor becomes available, it becomes of interest to use it as a probe to screen libraries in search of similar sequences. In the case of the ∂ opioid receptor, this search has taken on particular importance, for it seemed possible that the other types of opioid receptors defined pharmacologically, in particular the µ receptor, which is the major mediator of opioid antinociception, and the k receptor, might have similar nucleotide sequences. This has proved to be the case. Chen et al. (13) isolated a cDNA that expressed a receptor selective for µ ligands such as DAMGO, and which was negatively coupled to adenyl cyclase, while Yasuda et al. (14) isolated a clone that appears to code for the k receptor. Both of these receptors exhibit about 60% amino acid homology with the ∂ opioid receptor as well as with each other. The greatest homology, as expected, is in the seven putative transmembrane regions, while the least is in the N-terminal and C-terminal sequences, and in the second and third extracellular loops formed by amino acids between transmembrane regions 4 and 5 and 6 and 7. Of particular surprise and interest, however, is the very strong sequence homology--about 90%--of the third cytoplasmic loop, between transmembrane regions 5 and 6, in all three of these opioid receptor sequences. In other members of the G-protein-coupled superfamily, this region is thought to be critical in coupling to G-proteins, and consistent with the wide variety of G-proteins available for coupling to different receptors, the homology in this area is relatively low even between highly related receptors. For example, the somatostatin 1 and 2 receptors have about 40% sequence identity in this region, as do the b1 and b2-adrenergic receptors. The unexpectedly high homology in the case of opioid receptors suggest that these three receptors may couple to the same

G-proteins, and raises the possibility that there are other members of this opioid receptor family that differ in this region. Currently, investigators in several other laboratories are attempting to clone additional opioid receptors, including subtypes of μ, ∂ and k receptors, which have been identified on the basis of their pharmacology, as well as perhaps previously unknown types of opioid receptors.

Many other studies of opioid receptors are now under way making use of the available clones. Radioactive probes constructed from portions of the cDNAs are being used to examine the brain regional distribution of the message. Studies with the k opioid receptor cDNA demonstrate the presence of the message in hippocampus, hypothalamus, amygdala, and locus coeruleus (14) Similar studies are being carried out with μ and ∂ receptor sequences, as well as studies using antibodies made to portions of the opioid receptors' predicted amino acid sequences. The construction of mutant and chimeric receptors will be used to determine the role of particular regions of the receptors in ligand binding and function. The genomic DNA for opioid receptors is also being sequenced, and it is anticipated that the regulatory regions of the gene will soon be identified and explored.

**Purification, cDNA Cloning and Characterization of OBCAM**

Our own laboratory has taken a more traditional approach to identifying and characterizing opioid receptors. Several years ago, we reported purification of a 58 kD protein from bovine brain that bound opioid ligands in the presence of acidic lipids (15). Subsequently the cDNA for the protein was cloned, and found to have homology to immunoglobulin (Ig)-like proteins, particularly cell adhesion molecules (16). Thus it was called opioid binding cell adhesion molecule (OBCAM). OBCAM possesses three characteristic Ig domains, which are presumed to be oriented extracellularly. It has a single hydrophobic region at its C-terminus, but this is not long enough to traverse the cell membrane. Hence, it appears that it could not act by itself as a signal-transducing molecule, relaying ligand-binding events at the cell surface to functional molecules on the cytoplasmic side of the membrane. Furthermore, now that opioid receptors with homology to other G-protein-coupled receptors have been cloned, the question is raised of the relationship of OBCAM to these receptors. Does OBCAM in fact play any role in opioid receptor function?

We have several lines of evidence that it does. First, we have raised antibodies both to purified OBCAM and to peptides corresponding to portions of its predicted amino acid sequence, and demonstrated that these antibodies inhibit opioid binding to the purified protein as well as to brain membranes (17,18) One of the peptide antibodies, to a region of OBCAM known as MN-3, has been studied particularly intensively. By the use of fluorescent labelling together with confocal microscopy, our laboratory has shown that OBCAM-like material is not only present on the surface of NG108-15 cells, but is down-regulated by chronic treatment of these cells with opioid agonists, in a fashion parallel to the well-established down-regulation of opioid receptors in these cells (19) Thus down-regulation of the MN-3-material followed a similar time course, and was selective to d agonists. Other cell surface receptors were not affected, nor was an antibody to neural cell adhesion molecule (N-CAM) effective.

These data indicate that OBCAM plays a role in opioid receptor function, and may be closely associated with the receptor. The antibody data do not establish that OBCAM itself binds opioid ligands in situ, since an antibody to the molecule could alter its relationship to another molecule that was acting as receptor. In fact, the most direct way of testing OBCAM's ability to bind opioids would be to express it in a cell line, but this has not been possible. However, the cloning of OBCAM has made it possible to explore its function in another way, by selectively blocking its expression through the use of antisense cDNA. By transfecting NG108-15 cells with antisense cDNA to OBCAM, we have been able to create a stable cell line (ST7-3) in which opioid binding is greatly reduced relative to that of cells transfected with OBCAM sense (ST8-4), as well as non-transfected cells (20). The selectivity of this effect is suggested by the observation that binding of ligands to other cell surface receptors in ST7-3 cells was unaffected. Scatchard analysis of the binding indicated that most of the reduction was due to a decrease in receptor number, not affinity. Moreover, the remaining receptors could be further down-regulated by chronic opioid agonist treatment of the cells. Thus it appears that the OBCAM antisense has greatly reduced the number of opioid receptors on the cell, without affecting their intrinsic response to acute or chronic opioid agonist treatment. This result again is consistent with an association between OBCAM and another molecule functioning as the receptor.

The antisense OBCAM cDNA has also been used to create a line of transgenic mice. The antisense was injected into mouse oocytes, which were implanted into pseudo-pregnant females; upon birth and maturation, the mice were selectively bred to create a stable line of transgenics. Founder as well as first and second generation transgenic mice showed a reduced response to the antinociceptive effect of morphine, as determined by the tail flick test, and also reduced sensitivity to acute tolerance to morphine. In the latter test, the animals were pretreated with a fixed dose of morphine, followed by determination of the morphine AD50 in the presence of a fixed dose of the antagonist naloxone. Normal mice show an increased AD50 in the presence of naloxone, and the higher the pretreatment dose, the higher the AD50. This increase was greatly reduced in the transgenic mice.

These studies all suggest that OBCAM plays a role in opioid receptor function, but do not really address the question of how it could do so. Despite its original isolation on the basis of its opioid binding properties, the identification of G-protein-coupled opioid receptor clones suggests it does not play this role. Alternatively, however, it could regulate the coupling of the receptor to G-protein. As evidence for this, we have recently shown that coupling to G-proteins is also altered in ST7-3 (OBCAM antisense-transfected) cells (Gavitrapong et al., in press). To demonstrate this, we made use of the fact that cholera toxin (CTX) induces ADP-ribosylation of G-proteins only in the presence of ligand, which promotes coupling. In untransfected cells, or cells transfected with OBCAM sense, CTX was shown to induce, in the presence of opioid agonist, ADP-ribosylation of G-proteins that are identifiable on SDS gels as 39-41 kD material that is reactive to Gi2 and Go antibodies. In the ST7-3 cells, this labelling is greatly reduced, as was DADLE stimulation of GTPase, a standard measure of receptor-G-protein interaction, and DADLE inhibition of adenyl cyclase, to which opioid receptors are coupled in these cells. Guanine nucleotide inhibition of cyclase was also reduced. Thus in these cells, the presence of antisense OBCAM has somehow interfered with coupling of opioid receptors to G-proteins.

It is not clear how OBCAM, an extracellular molecule, could couple a seven transmembrane receptor to G-proteins, which are located on the cytoplasmic face of the cell membrane. In fact, to our knowledge there is no other evidence for the existence of a coupling molecule mediating between these two proteins, which are assumed by most investigators to interact directly. Perhaps it alters this interaction indirectly, by affecting the conformation of the seven-transmembrane receptor. Alternatively, since our studies in NG108-15 cells indicate that OBCAM can be down-regulated from the cell surface, the protein may be taken into the cell and interact directly with G-proteins.

## Conclusions

The cloning of the ∂ opioid receptor by two independent groups has been quickly followed by cloning of the other two major opioid receptor types, μ and k. It is likely that other opioid receptor types of subtypes will soon be discovered, and the availability of the receptor cDNAs will also allow investigators to explore their localization and function. Our own laboratory, following a more traditional approach, has isolated a protein that seems to play some role in opioid receptor function.

**Acknowledgment**
This work was supported by NIH-NIDA grants and by the Stark Fund of the Minnesota Medical Foundation.

## References

1. Pert C B, Snyder S H. Opiate receptor: its demonstration in nervous tissue. Science 179:1011-1014, 1973.
2. Simon E J, Hiller J M, Edelman J. Stereospecific binding of the potent narcotic analgesic (3H)etorphine to rat brain homogenate. Proc Natl Acad Sci USA 70:1947-1949, 1973.
3. Terenius L. Stereospecific interaction between narcotic analgesics and a synaptic plasma membrane fraction of rat cerebral cortex. Act Pharmacol 32:317-320, 1973.
4. Loh H H, Smith A P. Molecular characterization of opioid receptors. Ann Rev Pharmacol 30:123-147, 1990.
5. Evans C J, Kleith D E,Jr., Morrison K, Edwards R H. Cloning of a delta opioid receptor by functional expression. Science 258:1952-1955, 1992.
6. Kieffer B L, Befort K, Gaveriaux-Ruff C, Hirth C G. The ∂-opioid receptor: Isolation of a cDNA by expression cloning and pharmacological characterization. Proc Natl Acad Sci USA 89:12048-12052, 1992.
7. Koski G, Klee W A. Opiates inhibit adenylate cyclase by stimulating GTP hydrolysis. Proc Natl Acad Sci 78:4185-4189, 1981.
8. Childers S R, Snyder S H. Guanine nucleotides differentiate agonist and antagonist interactions with opiate receptors. Life Sci 33:759-762, 1978.
9. Blume A J. Interaction of ligands with the opiate receptors of brain membranes: Regulation by Ions and Nucleotides. Proc Natl Acad Sci USA 75:1713-1717, 1978.
10. Law P Y, Wu J, Koehler J E, Loh H H. Demonstration and characterization of opiate inhibition

of the striatal adenylate cyclase. J Neurochem 36:1834-1846, 1981.

11. Attali B, Saya D, Nah S Y, Vogel Z. K-opiate agonists inhibit Ca++ influx in rat spinal cord-dorsal root ganglion cultures. Involvement of a GTP-binding protein. J Biol Chem 264:347-353, 1989.
12. Dohlman H G, Caron M G, Lefkowitz R J. A family of receptors coupled to guanine nucleotide regulatory proteins. Biochem 26:2657-2664, 1987.
13. Chen Y, Mestek A, Liu J, Hurley J A, Yu L. Molecular cloning and functional expression of a μ-opioid receptor from rat brain. Mol Pharmacol 44:8-12, 1993.
14. Yasuda K, Raynor K, Kong H, Breder C D, Takeda J, Reisine T, Bell G I. Cloning and functional comparison of k and ∂ opioid receptors from mouse brain. Proc Natl Acad Sci USA 90:6736-6740, 1993.
15. Cho T M, Hasegawa J, Ge B L, Loh H H. Purification to apparent homogeneity to a μ-specific opioid receptor from rat brain. Proc Natl Acad Sci 83:4138-4142, 1986.
16. Schofield P R, McFarland K C, Hayflick J S, Wilcox J N, Cho T M Roy S, Lee N M, Loh H H, Seeburg P H. Molecular characterization of a new immunoglobulin superfamily protein with potential roles in opioid binding and cell contact. EMBO J 8:489-495, 1989.
17. Roy S, Zhu Y X, Lee N M, Loh H H. Different molecular weight forms of opioid receptors revealed by polyclonal antibodies. Biochem Biophys Res Comm 150:237-244, 1988.
18. Roy S, Zhu Y X, Loh H H, Lee N M. A monoclonal antibody that inhibits opioid binding to rat brain membranes. Biochem Biophys Res Comm 154:688-693, 1988.
19. Lane C M, Elde R, Lee N M, Loh H H. Immunofluorescence staining of an OBCAM-like protein resembles the ∂-opioid receptor in NG108-15 cells. Proc Natl Acad Sci USA 89:11234-11238, 1992.
20. Ann D K, Hasegawa J, Ko J L, Chen S T, Lee N M, Loh H H. Specific reduction of ∂-opioid receptor binding in transfected NG108-15 cells. J Biol Chem 267:7921-7926, 1992.

# 3

# Central Dopaminergic Mechanisms of Alcohol and Opiate Withdrawal Syndromes

**M. Diana, Z. L. Rossetti, G. L. Gessa**
"Bernard Brodie" Department of Neurosciences, University of Cagliari, Italy

**Summary**

Rats made dependent on ethanol show, upon suspension of treatment, a reduction of dopamine outflow in the nucleus accumbens, as measured by microdialysis, to which corresponds a lowered electrophysiological activity of mesolimbic dopaminergic neurons. Administration of ethanol increases biochemical and electrophysiological indices of dopaminergic function. Similarly, morphine dependent rats display a reduction of dopamine efflux and an abrupt abatement of dopaminergic neuronal activity in the mesolimbic system. As in the case of ethanol, administration of the suspended morphine increases neuronal function and its biochemical correlate in the terminal area. The results, irrespective of the substance abused, are interpreted as indicating a common substrate which becomes hypofunctional upon suspension of chronic administration of an addicting substance. These findings may be viewed as the neurophysiological correlate of the dysphoria associated to drug withdrawal in humans.

## Introduction

Drug withdrawal syndromes are an important phase in the evolution of drug dependence. Indeed, after protracted administration and abrupt suspension of ethanol and/or morphine, two factors come into play to reinstate drug taking behavior: 1) positive reinforcement due to the intrinsic properties of the drug and 2) negative reinforcement due to the absence of the drug itself. These two aspects may play a significant role in drug-seeking and consequently drug-taking behavior which eventually results in relapse.

The mesolimbic dopaminergic system is considered to play a crucial role in brain reward mechanisms (1) and in drug-induced reward (2). Accordingly, dopaminergic firing is increased by acute administration of both ethanol (3) and morphine (4). The increase in neuronal activity, in turn, is thought to be the cause of the augmented dopamine outflow in the nucleus accumbens observed with both drugs (5,6).

On the other hand, after prolonged treatment with ethanol and/or opioids the drug-withdrawal syndrome emerges. While the two syndromes are characterized behaviorally by a different symptomatology in each case, they share a profound dysphoric status which may be important in the subsequent relapse.

Recently, we found (7) that the ethanol withdrawal syndrome is accompanied by a reduction of neurophysiological and biochemical indices of mesolimbic dopaminergic activity. In a further effort to ascertain the mechanisms determining this condition we tested the possibility, suggested by others (8) that dopaminergic neurons are affected by depolarization inactivation and we studied the morphine withdrawal syndrome to verify the possibility of a lowered dopaminergic tone in this condition. Indeed, if the abatement of dopaminergic neuronal activity is responsible for the dysphoric status associated to drug-withdrawal, it must be expected in a similar condition irrespective of the abused substance.

## Materials and Methods

### *Subjects and Ethanol Treatment*

Male Sprague-Dawley rats (Charles River, Como, Italy) weighing 200-225 g at the beginning of treatment were used in all experiments. Ethanol (20% v/v) was administered intragastrically every 6 hours for 6 consecutive days at the dose of 2-5 g/kg. according to the method introduced by Majchrowitz (9). The dose was adjusted to the degree of ethanol intoxication, assessed by the experimenter, in order to maintain steady intoxicating levels of plasma ethanol (9). Briefly, before each ethanol administration, rats were evaluated and a score ranging from neutrality to coma was assigned. When neutrality was scored rats received the highest (5 g/kg) dose of ethanol. On the contrary, when severe signs of intoxication, such as loss of the righting reflex and coma, were scored no ethanol was administered in that session.

At intermediate levels of intoxication (sedation, ataxia 1,2,3) an intermediate ethanol dosage was administered (4,3,2,1 g/kg respectively).

### *Microdialysis Experiments*

On the morning of the sixth day of ethanol treatment (2 hours after last ethanol administration), subjects were anesthetized with chloral hydrate (200 mg/kg ip) and placed on a stereotaxic frame for fiber implantation as already described for the striatum (10). One dialysis fiber (AN69 HF, Hospal, France) per rat was inserted horizontally at the level of the nucleus accumbens at coordinates (A: 1.7; V: 7.3) from bregma according to the atlas of Paxinos and Watson (11). Fiber inlet and outlet were secured to the skull with dental cement and rats were then returned to their home cages to undergo the last day of ethanol treatment. On the seventh day (20-24 hours after last ethanol administration) rats were connected to the infusion pump and the dialysis membrane perfused (2 ml/min) with Ringer solution (NaCl 147 mM; KCl 3 mM; CaCl2 2.2 mM; MgSO4 1.2 mM; KH2PO4 0.4 mM; pH 7.2).

Twenty-minute perfusates (40 ml) were injected into the HPLC column (5mm Beckman ultrasphere C-18, 150 X 4 mm) for separation of DA, 3,4-dihydroxyphenylacetic acid (DOPAC) and homovanillic acid (HVA). Quantification was performed by electrochemical detection (Waters Amperometric detector M 460; elctrode voltage was set at 0.65 volts). The mobile phase consisted of citrate-acetate buffer 50 mM, pH4.6 containing octyl sulphate 30 mg/L, EDTA 0.1 mM and methanol 7% v/v.

*Electrophysiological Experiments*

Single cell extracellular recordings from antidromically identified meso-accumbens dopaminergic neurons were performed as already described for nigro-striatal DA cells (12) (see below). Briefly, 20-24 hours after last ethanol dose rats were tested for withdrawal signs and assigned a withdrawal score according to the rating scale of Lal et al. (13). The majority of animals (74%) chronically treated with ethanol displayed a robust withdrawal syndrome characterized by severe head, body and tail tremors, vocalization on handling, bracing posture, rigidity etc.. Each item was scored separately and the sum of the individual scores obtained was used to quantify the severity of the withdrawal syndrome. The rat was then placed on the stereotaxic frame for extracellular recording experiments.

*Subjects and Morphine Treatment*

Male Sprague-Dawley rats (Charles River, Como, Italy) weighing 200-225 g at the beginning of treatment were used in all experiments. Rats were housed four/cage with food and water available ad libitum under a 12 hours/12 hours light/dark cycle with lights on at 8.00 a.m.. Morphine hydrochloride was dissolved in sterile saline and administered twice daily (8.00 a.m./8.00 p.m.) for fourteen days according to the schedule of Acquas & Di Chiara (14). Briefly, the initial dose administered was 20 mg/kg and it was increased by 20 mg/kg every other day until the fourteenth day of treatment. Morphine doses up to 100 mg/kg were administered subcutaneously in a volume of 1 ml/kg whereas higher doses were administered intraperitoneally in a volume of 1 ml/ 0.1 kg.

Before electrophysiological recordings rats were tested for signs of opiate withdrawal such as: wet dog shakes, weight loss, diarrhoea, teeth chattering etc.

Subjects were temporarily anesthetized with halothane/room air inhalation anesthesia and the femoral vein was cannulated for intravenous administration of pharmacological agents. The trachea was then exposed and incised to allow tracheal intubation with a teflon catheter for artificial respiration. The paralyzing agent d-tubocurarine (4mg/kg) was administered intravenously and, once muscular paralysis was obtained, the rat was placed on a stereotaxic apparatus (Kopf) and the tracheal catheter connected to a mechanical rodent ventilator (7025 Stoelting) set to deliver 90 strokes/min (3 ml/stroke). All pressure and incision points were infiltrated with a long-lasting anesthetic (Xylocaine ointment 2%). All experiments were carried out in strict accordance with the Guiding principles in the care and use of animals approved by the Council of the

American Physiologic Society.

The skull was opened and a burr hole drilled above the ventro-tegmental area VTA (2.2-2.4 AP; 0.3-0.5 L, from lambda) according to Paxinos and Watson (11). A glass microelectrode (4-6 Mohms) filled with a 2% pontamine sky blue (PSB) solution in 0.5 M sodium acetate was lowered into the VTA. DA neurons were identified according to well-established electrophysiological features and by antidromic activation. Stimuli consisting of monophasic rectangular pulses (0.1-2.0 mA; 0.1-0.5 msec; 0.8 Hz) were generated by a Grass Instrument stimulator (S88) and stimulus isolation unit (SIU5) and delivered to the nucleus accumbens (AP 8.8; L 1.6 from lambda; V 7.1 from dura matter) (11) by means of a stainless steel Formvar coated, bipolar stimulating electrode (250 mm tip diameter). Stimulating current was monitored on the oscilloscope. At the end of each recording session the recording site was marked by iontophoretic ejection (1-10mA, negative current for 10 min) of PSB for histological verification. The extracellular neuronal signal from single neurons was amplified (Neurolog system) and displayed on a digital oscilloscope (Philips pm 3305) before storage on a magnetic tape for off-line analysis of the data. Firing rate and pattern analysis were performed as already described (15).

The number of spontaneously active DA neurons was determined by passing the electrode 8 times through 4 different areas within the VTA, according to the method proposed by Bunney & Grace (16). Coordinates of tracks were 2.2 and 2.4 mm anterior to lambda and 0.3 and 0.5 mm lateral to midline. The electrode was lowered twice in each spot while the total number of active cells were counted and divided by the number of passes to obtain cells/track.

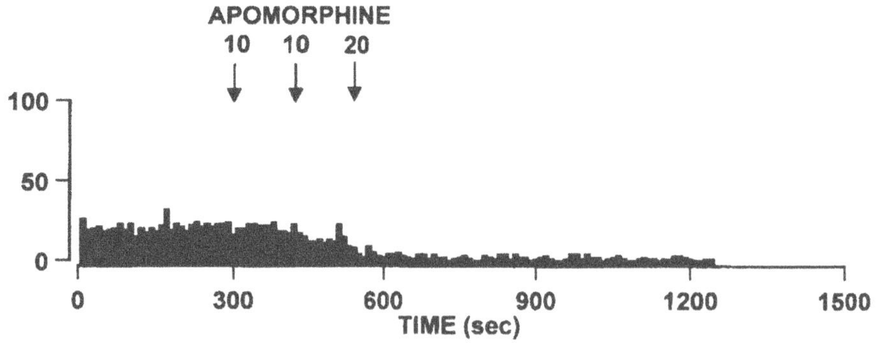


**Fig. 1.** The administration of apomorphine (10-40 mg/kg) iv to d-tubocurarine treated rats decreases firing rate

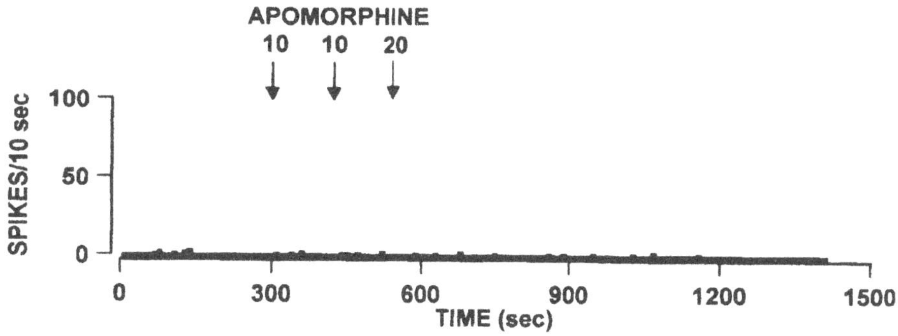


**Fig. 2.** In chloral hidrate anhaestetized rats apomorphine fails to alter the quiescent status of silent dopaminergic neurons

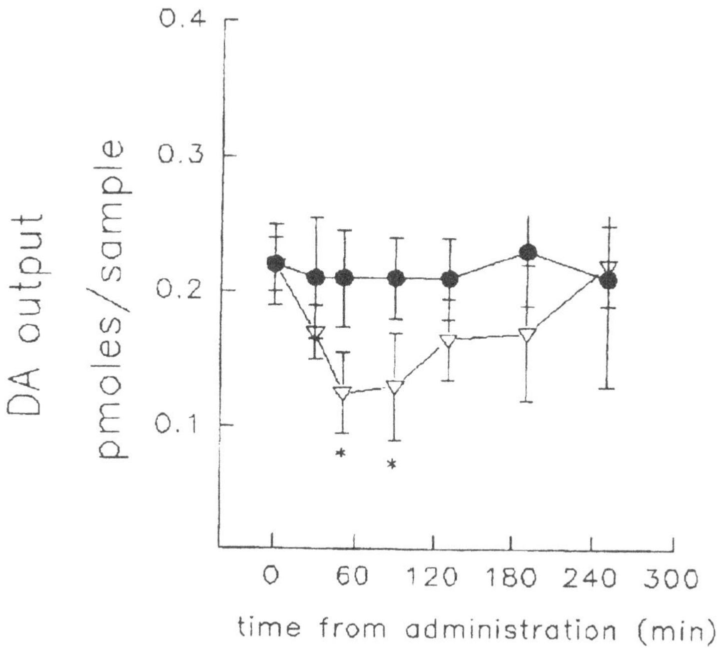


**Fig. 3.** The administration of Quinpirole reduces DA outflow in the nucleus accumbens

**Table 1.** Basal activity of mesoaccumbens neurons in morphine withdrawn rats

| | MW | C |
|---|---|---|
| N° of cells | 40 | 48 |
| cells/track | 0.96±0.16 | 1.20±0.26 |
| Firing rate (Hz) | 1.04±0.15** | 2.72±0.23 |
| Burst rate (burst/sec) | 0.16±0.05** | 0.46±0.10 |
| Spikes/burst | 2.05±0.20 | 2.45±0.15 |
| ARP (msec) | 1.78±0.11 | 1.55±0.07 |
| RRP (msec) | 3.22±0.16* | 2.69±0.11 |

The table illustrates various electrophysiological parameters recorded from morphine-withdrawn (MW) and saline treated rats (C). Cells/track= number of cells responding to antidromic identification from nucleus accumbens divided by the number of tracks. Firing rate= number of spikes per second (Hz). Burst rate= number of burst averaged for 1 second. Spikes/ burst= number of action potentials per each burst. ARP= Absolute Refractory Period. RRP= Relative Refractory Period.
*p<.05 **p<.01 Student's T-Test for unpaired observation

## Results

The possibility of a reduced incidence of spontaneously active mesolimbic dopaminergic neurons was tested in two groups of rats at 24 hours after suspension of ethanol treatment. The first group was immobilized with d-tubocurarine whereas the second group was anhaestetized with chloral hidrate. In chloral hidrate anhaestetized rats the cells/track was found reduced (0.22±0.06)

as compared with d-tubocurarine treated rats (1.04±0.27) (p<0.01 Student's t-test for unpaired observation). Administration of apomorphine (10-40 mg/kg) iv to d-tubocurarine treated rats decreased firing rate (Fig. 1).

In chloral hidrate anhaestetized rats, the effect of iv apomorphine was tested in 3 silent dopaminergic neurons. In these cases apomorphine (10-40 mg/kg) iv failed to alter the quiescent status of these neurons (Fig. 2). Further, the action of quinpirole (50 mg/kg) sc was tested (n=6) in microdialysis experiments. In this case, administration of quinpirole reduced DA outflow in the nucleus accumbens (Fig. 3).

The effect of morphine withdrawal on dopaminergic neuronal activity was studied in rats (n=5) which had received the last morphine administration 24 hours earlier. The control group (n=5) received an equal volume of saline 24 hours before recording began. Firing rates were found uniformly reduced across animals in the group of rats withdrawn from chronic administration of morphine. Indeed, of the 40 neurons recorded from this group only 6 were found to have frequencies of discharge above 2 Hz. In contrast, in the saline treated control group 29 out of 48 neurons fired with frequencies higher than 2 Hz. This reduction is not confined to the gross firing rate but it also includes the degree of absolute burst firing (Table 1). Quantitatively the diminution reaches values as low as about a third of control values for both parameters (Table 1). However, the RRP was found prolonged in DA neurons recorded from morphine withdrawn rats as compared with those observed in control animals (Table 1).

## Discussion

The present results indicate that the reduction in the cells/track index observed in anhestetized rats (8) and confirmed in this study, does not reflect a general finding of the ethanol withdrawal nor a depolarization inactivation of mesolimbic DA neurons in this condition. Indeed, the cells/track index is not reduced in d-tubocurarine immobilized rats and thus should be confined to the anhestetized preparation probably reflecting the interaction of the anhestetic agent with the alterations due to the ethanol-withdrawal syndrome. This contention is supported by two experimental facts: 1) apomorphine (unselective dopaminergic agonist) failed in increasing DA firing in d-tubocurarine immobilized rats. 2) apomorphine failed to restore firing in quiescent DA neurons recorded from chloral hydrate anhestetized rats. These effects should be expected if DA neurons were affected by depolarization block. Furthermore, administration of a different dopaminergic agonist such as quinpirole to freely moving rats, further decreased the already abated dopamine release as measured by the microdialysis probe in the nucleus accumbens. Again, even assuming that the neurons studied electrophysiologically would belong to a subpopulation not affected by depolarization block and thus unresponsive to the excitatory effects of apomorphine, an increment of DA release in the nucleus accumbens shoud have been observed with the dialysis method. The reduced cells/track index observed in chloral hydrate anhestetized rats, however remains to be explained.

In morphine withdrawn rats a reduction of firing rate, burst firing, but not cells/track was observed. These findings are reminiscent of those observed in ethanol-withdrawn rats (7) and corroborate the notion that central dopaminergic mechanisms of withdrawal may be similar and independent of the abused substance. This result points once again to the mesolimbic dopaminergic system as a pivotal pathway in the motivational properties of abused drugs (6). In the case of drug-withdrawal syndromes this role would be a reduction of its activity as opposed to its increment as a consequence of acute administration of addicting substances (6). The reduction in mesolimbic dopaminergic activity may suggest a link with the dysphoria observed in humans upon suspension of a chronic administration of addicting compounds. The drug-withdrawal induced dysphoria may, in turn, play a primary role in determining drug-seeking and drug-taking behavior (craving) thus determining relapse into drug addiction.

On a different perspective, therapeutic agents aimed at restoring "normal" dopaminergic functions in drug-withdrawn individuals may reduce the incidence of relapse. Accordingly, experimental studies in rats self-administering cocaine have shown that stimulation of D-3 dopamine receptors reduces self-administration behavior (17).

**Acknowledgements**

The authors wish to thank Mr. S. Aramo for his skilful technical assistance and Mr. Hugh Sugden for language editing of the manuscript.

**References**


1. Wise R A, Rompre P P. Brain dopamine and reward. Ann Rev Psychol 40:191-225, 1989.
2. Di Chiara G, Imperato A. Drugs abused by humans preferentially increase synaptic dopamine concentrations in the mesolimbic system of freely-moving rats. Proc Natl Acad Sci USA 85:5274-5278, 1988.
3. Gessa G L, Muntoni F, Collu M, Vargiu L, Mereu G P. Low doses of ethanol activate dopaminergic neurons in the ventral tegmental area. Brain Res 348:201-203, 1985.
4. Matthews R T, German D C. Electrophysiological evidence for excitation of rat ventral tegmental area dopamine neurons by morphine. Neuroscience 11:617-625, 1984.
5. Imperato A, Di Chiara G. Preferential stimulation of dopamine release in the nucleus accumbens of freely-moving rats by ethanol. J Pharmacol Exp Ther 239:219-238, 1986.
6. Di Chiara G, Imperato A. Opposite effects of m- and k-opiate agonists on dopamine release in the nucleus accumbens and in the dorsal caudate of freely moving rats. J Pharmacol Exp Ther 244:1067-1080, 1988.
7. Diana M, Pistis M, Carboni S, Gessa G L, Rossetti Z L. Profound decrement of mesolimbic dopaminergic neuronal activity during ethanol withdrawal syndrome in rats: electrophysiological and biochemical evidence. Proc Natl Acad Sci USA 90:5966-5969, 1993.
8. Shen R Y, Chiodo L A. Acute withdrawal after repeated ethanol treatment reduces the number of spontaneously active dopaminergic neurons in the ventral tegmental area. Brain Res 622:289-293, 1993.
9. Majchrowicz E. Induction of physical dependence upon ethanol and associated behavioral changes in rats. Psychopharmacologia 43:245-254, 1975.
10. Pani L, Gessa G L, Carboni S, Portas C M, Rossetti Z L. Brain dialysis and dopamine: does the extracellular concentrations of dopamine reflect synaptic release? Eur J Pharmacol 180:85-90, 1990.
11. Paxinos G, Watson C. The rat brain in stereotaxic coordinates, Academic Press, London, 1986.
12. Diana M, Young S J, Groves P M. Modulation of dopaminergic terminal excitability by D1 selective agents: further characterization. Neuroscience 42:441-449, 1991.
13. Lal H, Harris C M, Benjamin D, Springfield A C, Bhadra S, Emmett-Oglesby M W. Characterization of a pentylenetetrazol-like interoceptive stimulus produced by alcohol withdrawal. J Pharmacol Exp Ther 247:508-518, 1988.
14. Acquas E, Di Chiara G. Depression of mesolimbic dopamine transmission and sensitization to morphine during opiate abstinence. J Neurochem 58:1620-1625, 1992.
15. Diana M, Garcia-Munoz M, Richards J B, Freed C R. Electrophysiological analysis of dopamine cells from the substantia nigra pars compacta of circling rats. Exptl Brain Res 74:625-630, 1989.
16. Bunney B S, Grace A A. Acute and chronic haloperidol treatment: comparisons of effect on nigral dopaminergic activity. Life Sci 23:1715-1728, 1978.
17. Barak Caine S, Koob G F. Modulation of cocaine self-administration in the rat through D-3 dopamine receptors. Science 260:1814-1816, 1993.

# 4

# Endogenous Morphine and Codeine As Possible Physiological Ligands of μ-Opiate Receptors

**A. Tagliamonte**[1], **M. Guarna**[2], **E. Bianchi**[1]
[1] Institute of Pharmacology, University of Siena, Italy
[2] Institute of Histology and General Embryology, University of Siena, Italy

## Summary

Experiments carried out by indirect immunofluorescence and unlabelled antibody enzyme procedures revealed the presence of morphine-like immunoreactive material in the perikarya, fibers, and terminals of neurons in different, discrete areas of rat and human brain. The monoclonal and polyclonal anti-morphine antibodies used do not distinguish between morphine and codeine. Endogenous morphine seems to be stored in neurons as the 3-ethereal sulfate conjugate. This possibility is supported by the finding that, although active uptake of [3H]morphine has not been detected in brain synaptosomes, long-term i.c.v. injection of the tritiated opiate results in the accumulation of radioactivity inside the same neurons in which the endogenous alkaloids have been detected. Moreover, the sulfated compound showed a selective displacing effect toward [3H]DAMGO, although the affinity was lower than that of morphine HCl and DAMGO. Finally, morphine-3-ethereal sulfate presented the same poor inhibitory action of morphine HCl and DAMGO on rat striatal adenylate cyclase activity. Finally, striatal slices exposed to high K+ concentrations showed a rapid disappearance of the morphine-like immunoreactive material from neurons, indicating that endogenous alkaloids are released from neurons by depolarization.

## Introduction

Endogenous opiate alkaloids were first detected in mammalian tissue by radioimmunoassay (RIA) techniques in 1976 (8). Since then, two main laboratories, one directed by Spector at the Roche Institute in Nutley, New Jersey, and the other by Goldstein at the Addiction Research Foundation in Palo Alto, California, obtained a wealthy mass of information in this area.

Morphine and codeine have been partially purified in extracts of mammalian organs, included the CNS, then processed by reverse phase liquid chromatography, and detected by radioimmunoassay (RIA) based on the use of both polyclonal and monoclonal anti-morphine antibodies (10, 15).

The possible existence of endogenous opiate alkaloids in mammals met with a considerable skepticism by the scientific community. The most ordinary criticism raised was that of an artifact due to the detection of trace amounts of opiates that may result from contamination. Such criticism endured the determination of endogenous morphine and codeine by GCMS (5). The levels of codeine in the C.N.S. have been found consistently higher than those of morphine and, on a molar basis, both compounds show concentrations in the same order of magnitude of - and even higher than - those of met-enkephalin and ß-endorphin (22).

In order to eliminate the possibility of contamination a large-scale purification was carried out, not only by segregating equipment, reagents, and glassware, but by an elaborate system of tissue-less blank controls running through the entire procedure (like a pilot train) ahead of each brain work-up (24). Through such a procedure, not only morphine and codeine, but also 6-acetylmorphine was detected unequivocally in beef brain.

The administration of salutaridine and thebaine, the direct precursors of codeine in plants (13), produces in rats a rapid clear-cut increase of codeine and morphine concentrations both in brain and in peripheral organs (4). In addition, the administration of codeine induces a marked increase of CNS morphine concentration (4). Morphine and codeine levels have been found to increase by 2-3 times in the spinal cord of rats with chronic pain as produced by experimental arthritis (5). Moreover, it has been reported that prolonged food deprivation increases, by several times, the brain levels of endogenous morphine (15).

A real break-through in the field of endogenous opiate alkaloids came from the demonstration in mammal tissue of the synthesis of the skeleton of the morphine molecule. Both in vivo and in vitro studies in rats with [3H]reticuline, the direct precursor of salutaridine, showed the formation of the morphinan skeleton - i.e. [3H]salutaridine - in liver, but not in brain or other organs (18). It was concluded that in mammals is present the same biosynthetic pathway for opiate alkaloids as demonstrated in the poppy (24). Reticuline in the poppy derives from tetrahydropapaveroline (THP), a 16-carbon alkaloid produced in mammalian brain and liver homogenates from dopamine (2). The administration of L-DOPA to rats results in the synthesis of THP in brain (21). Interestingly, in urines collected from normal subjects and from parkinsonian patients receiving L-DOPA therapy, THP, codeine and morphine were identified and measured by GC/MS, using selected ion monitoring with deuterated internal standards (16). The urines of L-DOPA-treated parkinsonian patients showed THP and codeine concentration one order of magnitude higher that those of normal subjects, and some of the patients had a high urine content of morphine.

## Immunohistochemistry

Gintzler et al. (9) have reported evidence for intraneuronal localization of morphine-like immunoreactivity in mouse brain areas related to the vestibular, cerebellar and raphe systems. Immunohistochemical evidence for the presence of morphine-like material was also reported in chromaffin cells of the American eel (11). These data have been recently confirmed in rats by using a rabbit derived polyclonal anti-morphine antibody and a mouse-produced monoclonal one (1). The level of cross reactivity with codeine was 176% for the monoclonal and 100% for the polyclonal antibody, respectively. Two goat immunoglobulins, an anti-rabbit for the polyclonal component and an anti-mouse for the monoclonal one, conjugated with fluorescein isothiocyanate, were used as secondary antibodies. For the peroxidase anti-peroxidase (PAP) detection a goat anti-rabbit immunoglobulin was used as a secondary antibody.

The antisera revealed numerous nerve cell bodies with a variable morphology throughout the cerebral cortex; many dendrites and nerve fibers were identified in the more superficial layers. Typical small pyramidal cells were visible in some cases. In the caudate-putamen numerous medium-size immunoreactive perikarya were visible (Fig. 1). Nerve cell bodies and fibers were stained at all levels in the hippocampal formation. Dot-like immunoreactivity was observed around granule cells of the dentate gyrus and pyramidal cells of the dorsal hippocampus (Fig. 2). Furthermore, immunoreactive nerve cell bodies could be revealed in the hypothalamus, in the metathalamic area and in the brain-stem.

In the cerebellum numerous granule cells in the granular layer and small neurons and fibers in the molecular layer appeared to be immunoreactive. Immunostained cell bodies were also visible in the more superficial laminae of the dorsal horn and around the central canal of the spinal cord. In the anterior horn immunoreactive nerve fibers and terminals were observed around the perikarya of some motor neurons. Immunoreactivity was situated inside the cell body or, sometimes, in dots around the perikaryon in a ring-like fashion.

Morphine immunoreactive neurons of different rat brain areas were also studied by electron microscopy with the peroxidase-antiperoxidase immunocytochemical technique. Immunoreactivity was localized in neuronal perikarya, dendrites and nerve terminals, which sometimes appeared in direct apposition to unlabelled perikarya or dendrites (Fig. 3).

## Morphine Storage

According to Donnerer et al. (5) endogenous morphine in brain is stored as a conjugate. In fact, the original extraction procedure with acidified methanol gave a final concentration of free morphine much lower than that obtained when tissues were exposed to strong acid and boiling. Tissue homogenates were to added concentrated HCl up to a final 10% concentration, and heated at 100° C for 30 min. Such a treatment is effective in hydrolyzing morphine sulfate and glucuronide esters as well as disrupting binding to proteins, yet is to weak to cause rearrangement of the morphine molecule. The described procedure increased the extraction of free morphine, but did not modify that of codeine. Moreover, incubating tissues with the enzyme arylsulfatase increased free morphine by the same extent as after boiling in acid, whereas ß-glucuronidase had no such effect. Since arylsulfatase specifically hydrolyzes morphine-3-sulfate but not morphine-6-sulfate, the conclusion was that endogenous morphine is stored in tissues as 3-ethereal sulfate. It is relevant that this compound is 3 times more active than morphine in analgesia tests (17, 25). The subcellular distribution indicated that both

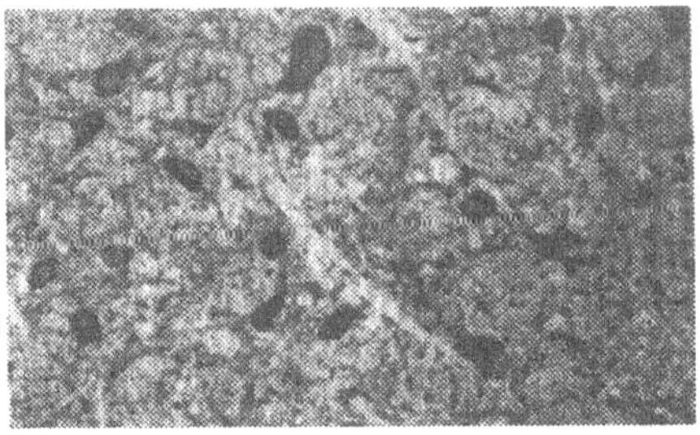

**Fig. 1.** PAP immunostaining of morphine in rat brain: coronal sections of caudate-putamen: numerous immunostained perikarya with variable morphology are visible
Scale: 30 μm

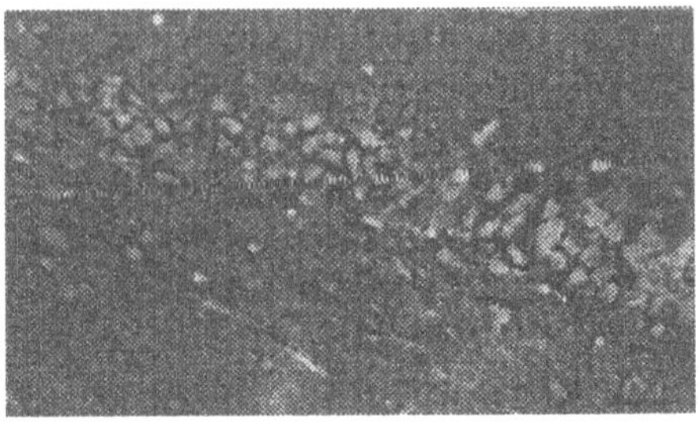

**Fig. 2.** PAP immunostaining of morphine in rat brain: coronal sections of dorsal hippocampal formation. The perikarya of hippocampal neurons appear surrounded by immunoreactive dot-like structures, resembling nerve terminals
Scale: 50 μm

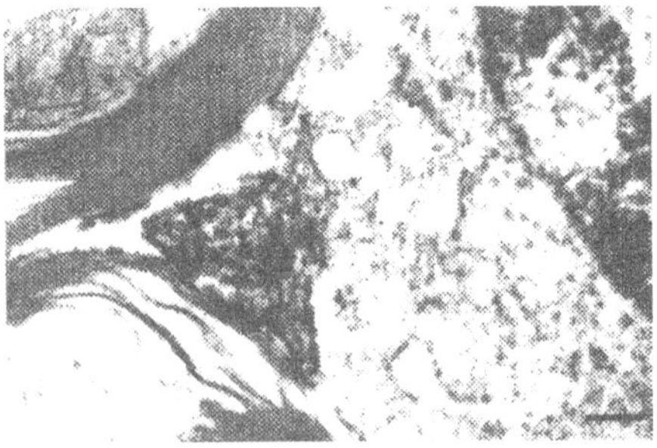

**Fig. 3.** Electron-micrograph of a morphine positive terminal in rat hippocampal molecular layer. PAP immunostaining. The nerve terminal forms probable synaptic contact with the soma of an unlabelled cell
Scale: 0.15 μm

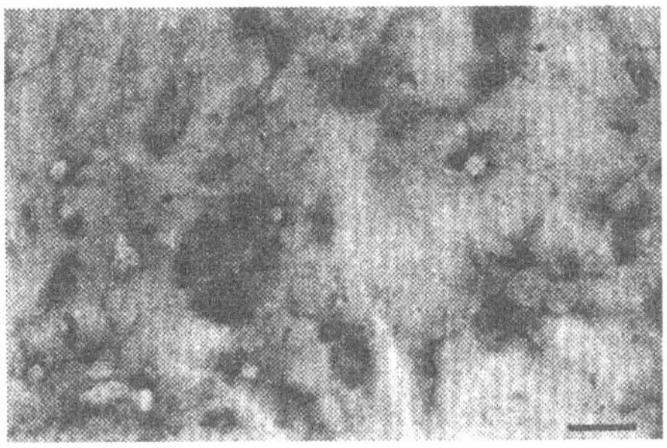

**Fig. 4.** Microautoradiography of rat brain-stem and hippocampus. Numerous silver grains are localized inside the cell body of brain-stem neurons, and pyramidal cells of hippocampus
Scale: 45 μm

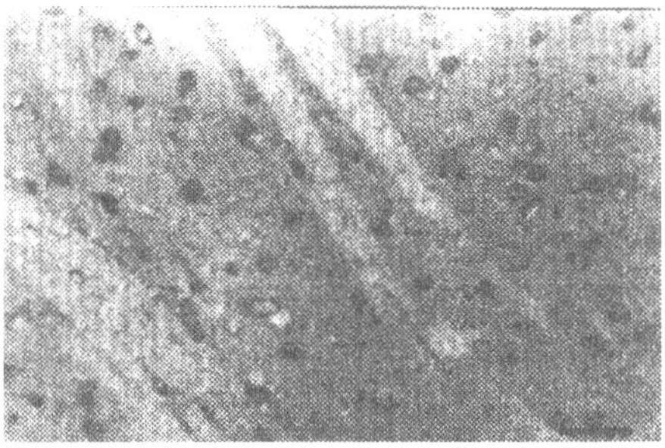

**Figs. 5-6.** Sections from striatal slices incubated in the presence of 5 mM K+ and 50 mM K+ Ringer solution. Numerous strongly immunoreactive neurons are visible in (5), whereas the immunostaining is dramatically decreased in (6)
Scales: 25 μm

codeine and morphine are mostly concentrated in the synaptosomal fraction, but are also present in the nuclear and microsomal fraction (5).

The presence of morphine and codeine in brain areas rich of specific μ-opiate binding sites raises the question as to the possible role of these alkaloids as neurotransmitters or neuromodulators. Among the criteria necessary to meet the definition of neurotransmitter is the existence of a specific mechanism for removing the candidate compound from the synaptic cleft (18). Thus, a possible [3H]morphine uptake by synaptosomes of different rat brain areas was studied. Briefly, a crude mitochondrial-synaptosomal fraction (P2) was incubated with [3H]morphine 20 nM (New England Nuclear, 83 Ci/mmol) at 37°C in Krebs-Henseleit buffer in the presence of 0.1 mM naloxone (in order to exclude the binding to opiate receptors), for 20, 40, and 60 minutes respectively. Brain synaptosomes from mid-brain, cerebral cortex and cerebellum failed to take-up and store any trace of the tritiated alkaloid. The same experiments were repeated in the presence of [3H]morphine-3-sulphate (New England Nuclear, 83 Ci/mmol). During incubation periods of 1-60 minutes duration, at 37° C., rat brain synaptosomes failed to show any concentrating activity of the sulfated alkaloid.

On the other hand, the slow i.c.v. infusion of [3H]morphine demonstrated the capacity of specific brain neurons to store the alkaloid, as detected by microautoradiographic techniques. Three male rats were chronically inserted a stainless-steel cannula in the skull, leading to the lateral ventriculi, and a subcutaneous osmotic minipump, releasing 1.3 mCi of [3H]morphine in 30 ml of saline/day/7 days, was connected to the cannula. Twenty hours after the end of infusion (i. e., when more than 5 morphine half-lives had elapsed), animals were anesthetized, sacrificed and their brain processed for microautoradiography (14). This technique showed the presence of numerous silver grains strictly localized inside the same cell types and areas as observed by immunocytochemistry (1) (Fig. 4). No signal of radioactivity was present outside the neurons. Inasmuch as it was demonstrated that morphine in brain is present as the 3-ethereal sulfate derivative, the simplest explanation of [3H]morphine accumulation inside the neurons could be that of an initial passive diffusion, followed by biotransformation through conjugation before storage.

**Table 1.** Displacing effect of different opioids on [³H]DAMGO, [³H]DPDPE and [³H]U 69-593 binding

| OPIOID | IC50 [³H]DAMGO (rat) | [³H]DPDPE | [³H]U69-593 (guinea-pig) |
|---|---|---|---|
| Morphine HCl | 0.19±0.04 | 0.63±0.07 | 0.50±0.11 |
| Morphine-3-ethereal sulfate | 6.00±0.78 | > 1.000 | 70±16 |
| DAMGO | 0.047±0.006 | 3.10±0.58 | 5.6±0.8 |
| DPDPE | 6.00±0.86 | 0.034±0.006 | 5.7±0.7 |
| U 50-488 | 1.95±0.12 | 30±4 | 0.017±0.003 |

## Morphine-3-Sulfate Activity

In order to clarify the biological activity of morphine-3-ethereal sulfate, the conjugated derivative (synthesized according to Mori et al.10, and kindly supplied by Dr. G. Cignarella, Pharmaceutical Chemistry Department, University of Milan) was compared to morphine HCl and other opioids for its capacity of displacing the binding of specific ligands from μ-, ∂- and k-receptors, and for that of inhibiting adenyl cyclase activity in rat striatum.

Table 1 shows the IC50 of morphine-3-sulfate on the binding of [3H]DAMGO, [3H]DPDPE and [3H]U 69-593, as compared to that of morphine HCl, DAMGO, DPDPE and U 50-488. The conjugated compound showed a selective displacing effect toward [3H]DAMGO, with an affinity from one to two orders of magnitude lower than that of both morphine HCl and DAMGO.

Table 2 compares the inhibitory effect of morphine-3-ethereal sulfate on striatal adenyl cyclase activity to that of morphine HCl, DAMGO and DPDPE. The conjugated derivative presented an inhibitory action similar to that of morphine HCl and DAMGO,

**Table 2.** Effects of different opioids on basal adenylate cyclase activity in rat striatal membranes. Values are mean ±s of 4 experiments

| OPIOID | Concentration 1μM | 10μM | 100μM | 1mM |
|---|---|---|---|---|
| | % inhibition | | | |
| Morphine HCl | 8±2 | 12±3 | 16±3 | 20±2 |
| Morphine-3-ethereal sulfate | 6±3 | 13±1 | 16±4 | - |
| DAMGO | 9±3 | 15±2 | 16±3 | 20±3 |
| DPDPE | 19±4 | 21±3 | 22±4 | - |

and lower than that of DPDPE.

**Morphine Release**

Neurotransmitters and neuromodulators are released from neurons by depolarization in a $Ca^{2+}$ dependent manner. Therefore, the demonstration of $Ca^{2+}$ dependent release from brain tissue constitutes one of the most direct criteria for the role of a compound as that of a neurotransmitter. Morphine and its congeners can be detected with different highly specific methods. However, unfortunately the sensitivity of all these methods (in the order of nmoles) is far below the concentration that the alkaloid may theoretically reach in the incubation medium in an in vitro release experiment or inside a microdialysis cannula. Thus, it was decided to use an indirect approach, namely the effect of a high (50 mM) $K^+$ concentration on the possible release of morphine in discrete brain areas. An entire caudate-putamen was dissected from a rat brain and incubated for 10 minutes at 37° C (millimolar Ringer composition: NaCl: 118.3; KCl: 4.7; $MgSO_4$: 1.2; $KH_2PO_4$: 1.2; $CaCl_2$: 2.5; $NaHCO_3$: 25; Glucose: 1.1; total volume: 10 ml) in a glass perfusion chamber. The tissue was continuously aerated with 95% $O_2$ and 5% $CO_2$. The second caudate-putamen from the same animal was incubated for 10 minutes with a Ringer solution containing 50 mM KCl, to replace an equimolar amount of NaCl.

The two striata were then fixed in paraformaldehyde. Sections of 50 mm prepared by a tissue chopper were processed by the PAP method as described above. The effects of high $K^+$ concentration were particularly clear-cut in the most superficial slices of the nucleus. Figures 5-6 show that after 10 minutes of depolarization the immunostaining of morphine-like material was dramatically decreased in striatal neurons.

**Discussion**

Experiments carried out with indirect immunofluorescence and unlabelled antibody enzyme procedures allowed to determine the presence of morphine-like compounds inside the perikarya, fibers, and terminals of neurons in different and discrete rat brain areas, as well as in several brain areas of humans. The anti-morphine antibodies used have a high cross-reactivity with codeine, so that the images obtained do not permit to separately map the distribution of the two alkaloids. Indeed, since in the rat's CNS the morphine concentration is about half of that of codeine, the question of distinct functional roles of the two compounds cannot be answered.

We failed to observe an active uptake of both [3H]morphine and [3H]morphine-3-ethereal sulfate by rat synaptosomes prepared from different brain areas. However, some neurotransmitters, e. g. acetylcholine and most neuropeptides, are not taken up by nerve terminals. Therefore, while the demonstration of a specific uptake mechanism would speak in favor of a neurotransmitter role of morphine, its absence does not necessarily speaks against it. A significant observation is the disappearance of the morphine-like immunoreactive material in striatal slices exposed to high $K^+$ concentration, which clearly indicates that endogenous alkaloids are released from neurons by depolarization.

According to Donnerer et al. (5) endogenous morphine is stored in the brain as a conjugate, most likely as 3-ethereal sulfate, a compound with a low affinity and high selectivity for μ-receptors, and an efficacy higher than that of morphine itself (17, 25). The conjugated compound showed a selective displacing effect toward [3H]DAMGO, with an affinity from one to two orders of magnitude lower than that of both morphine HCl and DAMGO. Finally, morphine-3-ethereal sulfate presented the same poor inhibitory action of morphine HCl and DAMGO on rat striatal adenylate cyclase activity. Moreover, no immunostaining reaction was demonstrated with anti-morphine antibodies in sections pre-incubated the presence of morphine 3-ethereal sulfate (1). Thus, a reasonable explanation of the accumulation of radioactivity observed in neurons following long-term i.c.v. infusion of [3H]morphine, could be that of initial passive diffusion, followed by biotransformation through sulfate conjugation, and storage in granules.

The detection of endogenous morphine-like compounds within neurons by immunocytochemistry, the accumulation of [3H]morphine by neurons, and the indirect evidence of a depolarization-dependent release, strongly support the hypothesis that codeine and/ or morphine may be physiological ligands for the m receptors in the brain.

It may be relevant to note that in the recently cloned d and m opiate receptors an aspartic acid residue is present in the third membrane-spanning hydrophobic domain (3, 6, 12), in a position, 113, considered crucial for the interaction of various ligands with a protonated amine group. The importance of such a residue has been demonstrated in a series of experiments that combine site-directed mutagenesis with pharmacophore analysis (20). When another amino acid is substituted for this aspartic residue, catecholamines (19) and acetylcholine (7) are no longer effective agonists for adrenergic and muscarinic receptors, respectively. Thus, it would be of interest to consider if the aspartic acid residue 113, present in both m and d opiate receptors, has the same relevance for the binding of peptide and alkaloid ligands. The detection of a possible difference could help in clarifying the recent finding (23) that a short-term treatment of NG108-15 cells with an alkaloid or opioid peptide agonist results in a rapid (45 minutes), marked (70 %), and dose dependent decrease in the binding of both agonist and antagonist peptides, with no effect on the binding of opiate alkaloids. Such an effect is observed in intact cells or in membrane preparations; thus excluding a mechanism of receptor internalization (23). Beyond any doubt about the possible molecular mechanisms involved, such a peptide-selective binding change could find a physiological explanation in the existence of a distinct role in the activation of opioid receptors by endogenous alkaloids and peptides.

## References


1. Bianchi E, Alessandrini C, Guarna M, Tagliamonte A. Endogenous codeine and morphine are stored in specific brain neurons. Brain Res 627:210-215, 1993.
2. Battersby A R, Binks R, Francis R J, McCaldin D J, Ramuz H. Alkaloid biosynthesis. 1-Benzylisoquinolines as precursors of thebaine, codeine and morphine. J Chem Soc Part IV:3600-3610, 1964.
3. Chen Y, Mestek A, Liu J, Yu L. Molecular cloning and functional expression of a μ-opioid receptor from rat brain. Brain Res 44:8-12, 1993.

4. Donnerer J, Cardinale G, Coffey J, Lisek C A, Jardine I, Spector S. Chemical characterization and regulation of endogenous morphine and codeine in the rat. J Pharmacol Exp Ther 242:583-587, 1987.
5. Donnerer J, Oka K, Brossi A, Rice K C, Spector S. Presence and formation of codeine and morphine in the rat. Proc Natl Acad Sci USA 83:4566-4567, 1986.
6. Evans C J, Keith D E Jr, Morrison H, Magendso K, Edwards R H. Cloning of a delta opioid receptor by functional expression. Science 258:1952-1955, 1992.
7. Fraser C M, Wang C D, Robinson D A, Gocayne J D, Venter C. Site-directed mutagenesis of M1 muscarinic acetylcholine receptors: conserved aspartic acids play important roles in receptor function. Mol Pharmacol 36:840-847, 1989.
8. Ginzler A R, Levy A, Spector S. Antibodies as a means of isolating and characterizing biologically active substances: presence of a non-peptide, morphine-like compound in the central nervous system. Proc Natl Acad Sci USA 73:2132-2136, 1976.
9. Ginzler A R, Gershon M D, Spector S. A non-peptide morphine-like compound: immunocytochemical in the mouse brain. Science 199:447-448, 1978.
10. Goldstein A, Barrett W, James F I, Lowney L I, Weitz C J, Knipmeyer L L, Rapaport H. Morphine and other opiates from beef brain and adrenal. Proc Natl Acad Sci USA 82:5203-5207, 1985.
11. Hathaway C B, Epple A. Catecholamines, opioid peptides, and true opiates in chromaffin cells of the eel: immunohistochemical evidence. Gen Comp Endocrinol 79:393-405, 1990.
12. Kieffer B L, Befort K, Gaveriaux-Ruff C, Hirt C G. The ∂-opioid receptor: isolation of a cDNA by expression cloning and pharmacological characterization. Proc Natl Acad Sci USA 89:12048-12052, 1992.
13. Kirby G W. Biosynthesis of the morphine alkaloids. Science 155:170-173, 1967.
14. Knoche H W. Radioisotopic Methods for Biological and Medical Research, Oxford University, New York, 1991.
15. Lee C S, Spector S. Changes of endogenous morphine and codeine contents in the fasting rats. J Pharmacol Exp Ther 257:647-650, 1991.
16. Matsubara K, Fukushima S, Akane A, Kobayashi S, Shiono H. Increased urinary morphine, codeine and tetrahydropapaveroline in parkinsonian patients undergoing 1-3,3-dihydroxyphenylalanine therapy: a possible biosyntethic pathway of morphine from 1-3,4-dihydroxyphenylalanine in humans. J Pharmacol Exp Ther 260:974-978, 1992.
17. Mulder G J. Pharmacological effects of drug conjugates: is morphine 6-glucuronide an exception? TIPS 13:302-304, 1992.
18. Schwartz J H. Chemical basis of synaptic transmission. In: Principles of neural Science, edited by Kandel E R, Schwartz J H. Elsevier, North Holland, 1981, pp 106-120.
19. Strader C D, Gaffney T, Sugg E E, Candelore M R, Keys R. Allele-specific activation of genetically engineered receptors. J Biol Chem 266:5-8, 1991.
20. Turner A, Baker K, Algeri S, Frigerio A, Garattini S. Tetrahydropapaveroline: formation in vivo and in vitro in rat brain. Life Sci 14:2247-2257, 1974.
21. Verhoef J, Wiegant V M, De Wield D. Regional distribution of $\alpha$- and $\gamma$-type endorphins in rat brain. Europ J Pharmacol 231:454-460, 1982.
22. Von Zastrow M, Keith D E, Evans C J. Agonist-induced state of the ∂-opioid receptor that discriminates between opioid peptides and opiate alkaloids. Brain Res 44:166-172, 1993.
23. Weitz C J, Faull K F, Goldstein A. Synthesis of the skeleton of the morphine molecule by mammalian liver. Nature 330:674-677, 1987.
24. Yaksh T L, Harty G J. Pharmacology of the allodynia in rats evoked by high dose intrathecal morphine. J Pharmacol Exp Ther 244:501-507, 1988.

# 5

# Drugs of Abuse Craving in Free-Choice Experimental Conditions

**M. Sala, D. Braida, M. Colombo, M.P. Leone, E. Gori**
Institute of Pharmacology, Faculty of Mathematical, Physical and Natural Sciences, University of Milan, Italy

## Summary

Given that the techniques used to test drug abuse liability are not free from criticism, a series of oral free-choice experimental procedures was adopted. Firstly, to rats water-deprived for 23h, a choice between pure and fentanyl citrate (5μg/ml) or buprenorphine (25 μg/ml) or etonitazene HCl (5μg/ml), a very potent μ-opiate, containing aspartame solution, adopted as very potent sweetener, was daily offered. Even if behavioural effects were evident and exhibited tolerance and dependence, no preference was shown. In a second procedure, when the same free-choice was allowed for 24h, no evident behavioural effects and no preference for etonitazene or a stimulant drug, cocaine HCL (300 μg/ml), were detected. However, it was possible to select a group of etonitazene preferring rats. When etonitazene was offered in a free-choice only 1 day over 3, rats exhibited preference versus etonitazene even if evident behavioural effects were absent. Finally, when a gustatory marker (100 μg/ml) was introduced into one of the two solutions, using 1 or 24h schedule, and adopting etonitazene or morphine (500 μg/ml), preference was always shown for the less bitter solution.

## Introduction

Although the evidence is only anecdotal, it is generally accepted that many wild animals sometimes show a craving for vegetables containing drugs of abuse (1). Furthermore over the last 20 years, it has also been shown that both primates and rodents are capable of intravenously self-administering most of the drugs of abuse in certain experimental settings (2). However, there are many, and frequently under-rated, objections as to the validity of the experimental procedures used in such cases: it is doubtful that some important drugs of abuse, such as LSD, mescaline (3) and nicotine (4), are really self-administered; the means of administration is unnatural; due account is not taken as to the stress induced by confinement and laboratory environment; there is a lack of social inhibition and competing reinforcements and the drug is constantly available at no cost. Oral consumption is certainly more natural, easy to organize and allows free-choice, but its experimental use is hampered by two factors: many of the drug solutions are spontaneously refused because of their inherent aversive properties, and the level of their consumption is so low that no evident behavioral effect can be observed; secondly, at least in rodents, any delay in the eliciting of effects could theoretically interfere with the temporal association of assumption and reward.

The schedule-induced polydipsia method (5) rather similar to self-administration, has been widely used to induce excessive ingestion of opiates (6), ethanol (7), cocaine (8) and benzodiazepines (9). On the other hand the drinking procedure in which animals are offered the drug in their home cage , even if a more physiological method, has been previously restricted to alcohol (10) and contradictory data exist concerning the consumption of narcotics in monkeys (11) and rats (12,13,14). In order to circumvent these disvantages, we decided to use, in a two-bottle drinking test, only those opiates that are so highly active by the oral route that their effects can be observed even when they are taken in very low, and therefore non-aversive concentrations. In addition, we adopted the use of potent synthetic sweeteners such as aspartame.

## Experiment 1: 1h Free-Choice Procedure

*Animals*

Male Wistar rats, initially weighing 180-300 g lodged in a climatized room (humidity 50±5%; temperature 22±2°C; 12 h light) were individually housed in standard laboratory cages.

Procedure

Since our previous findings (15) using schedule-induced polydipsia suggested at least for opiates, that drug preference is not affected by premedication, and being the polydipsic schedule rather complex, we decided to adopt the equally classical but more natural two-bottle drinking test without premedication.

This procedure, originally developed by Myers (16) for alcohol, was used as previously described (17). Briefly, male Wistar rats left in their home cages to drink for only 60 min a day from two graduated bottles, whose positions were changed every day.

One bottle contained a pure Asp solution (Pierrel, 0.18%), and the other an Asp solution with the addition of fentanyl citrate (F) (5μg/ml).

At the end of every drinking session an assessment was made of fluid intake (ml/kg), analgesia, tested by the tail-flick method (18), catatonia evaluated according to Dunstan et al.(19) with a maximum score of 6, and jumpings according to Huang et al. (20).

## Results

As shown in Fig.1, when the rats were offered a choice between pure and F-containing Asp, they showed no preference over a period of 14 days, despite the fact that the clear initial analgesic effect gave way to tolerance within few days.

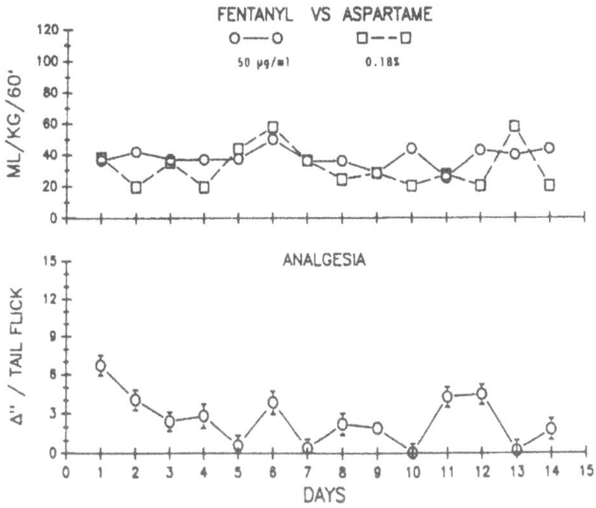


**Fig. 1.** Rats offered a choice between pure and F-containing Aspartame show no preference over a period of 14 days, despite the fact that the clear initial analgesic effect gave way to tolerance within few days

## Experiment 2: 1h Free-Choice Procedure and Withdrawal

Since no preference was shown in experiment 1, we decided not only to test other opiates, but also to evaluate the possible role of withdrawal. Thus buprenorphine (B) (25 μg/ml) or Etonitazene HCl (E) (5 μg/ml),a synthetic μ-opiate reported to be 1000 times more potent than morphine by the parenteral route (21), were simultaneously offered with a pure Asp solution in a free-choice situation for 35 days. 24 h the last drinking session a withdrawal syndrome was precipitated by naloxone (5 mg/kg/ip) and scored according to Grant and Redmond (22). Three days after, a free-choice was again allowed over ten days. After this period, a new precipitated withdrawal syndrome was carried out followed, 1h after, by a readmission to the free-choice drinking test over 1 week. Also in this case an assessment was made of fluid intake (ml/kg) and behavior as above.

## Results

When the rats were offered a choice between pure and B or E containing Asp, they showed no preference over a period of 35 days, despite the fact that the initially clearly evident analgesia had slowly decreased to nil by the 15th day (data not shown). Once again, the presence of clear naloxone-induced withdrawal symptoms did not affect preference, either three days (total score= 10 for B and 11 for E) or 1 hour (total score=9 for both drugs) after precipitated withdrawal. However, during this period the amount of ingested drug induced a very slight analgesia.

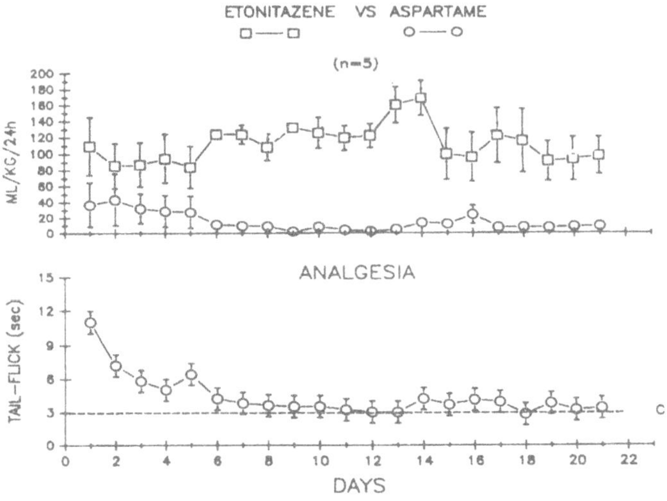


**Fig. 2.** 5 rats out of 11 have a significant preference for Etonitazene when a choice is offered between Etonitazene and Aspartame

### Experiment 3: 24 h Free-Choice Procedure

It is possible to make a number of objections to previous experiments: the drug contact time was too short; given the deprivation of water, the animals drive to satisfy their thirst is such that no discrimination is made between the offered solutions; and the fact that there is a delay between drug consumption and the beginning of behavioral effects.

To circumvent these objections, we decided to offer the choice between pure and Asp containing E, using a continuous 24 h a day schedule. Since the neurodepressant effects of opiates may interfere with drinking ability, we tested cocaine HCl (C) (300 µg/ml), a stimulant which, contrary to common opinion, is also active when taken orally. The behavioral observations, evaluated as above, were made every early morning at 9.00 a.m. since rodents are nocturnal and preferentially drink during the dark cycle. For C, the typical pattern of stereotypy, was tested by counting the number of rearings within 2 minutes.

### Results

Once again the animals seemed to show no preference between the pure and E containing Asp solutions (data not shown). The mean amount of the ingested drug solution was only slightly higher, and the elicited analgesia was very low. Nevertheless, the abstinence score was still 10. However, ANOVA test of the amounts of the offered solution ingested by each rat showed that 5 out of 11 had a significant preference for E: on average, they drank five times more of the drug than the pure Asp solution (Fig.2). As expected, this group of rats reflected a strong analgesic effect with subsequent tolerance. When C was offered for 24 hours the amount ingested was obviously less because of its perceived bitter taste (Fig. 3). Nevertheless, this amount remained constant even though rearings were evident and showed progressive sensitization.

Since it is generally agreed that it is quite difficult to demonstrate the reinforcing efficacy of drugs when these are continuously available, we carried out preliminary experiments using and alternating schedule in which only the pure Asp solution is offered for two days, followed the day after by a simultaneous choice between E and the Asp solution.

## Results

Regarding the findings of the alternate schedule, all of the rats, after the first 3 free-choice sessions, nearly always choose the drug solution even if the position of the bottle was changed from the right to the left. The amount of ingested fluid containing E was double in comparison to Asp fluid. A slight analgesic effect and a mild withdrawal syndrome (total score=8) was observed (data not shown).

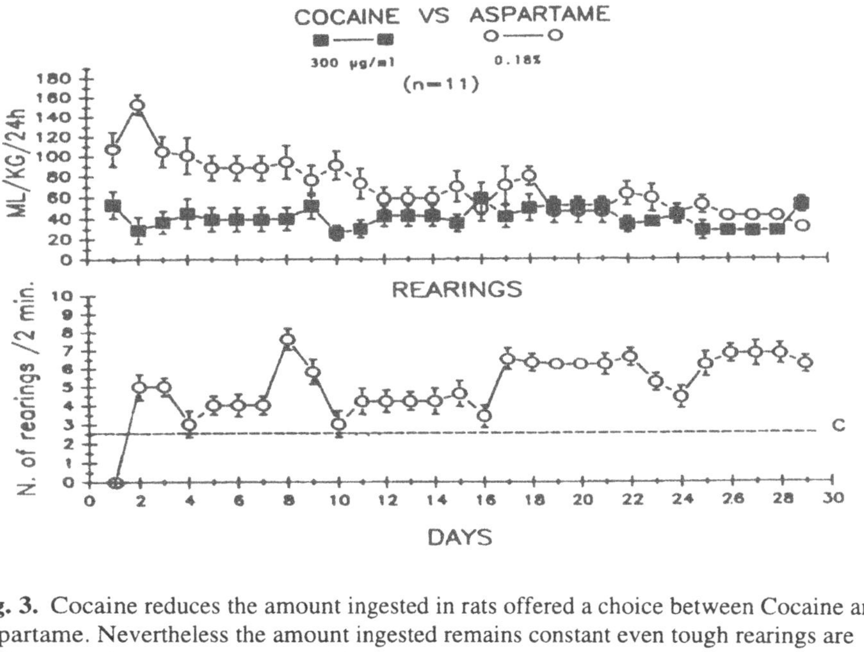


**Fig. 3.** Cocaine reduces the amount ingested in rats offered a choice between Cocaine and Aspartame. Nevertheless the amount ingested remains constant even tough rearings are evident and show progressive sensitization

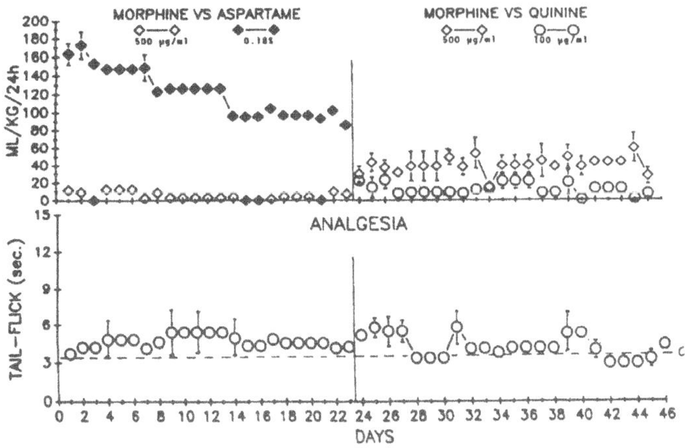


**Fig. 4.** In rats submitted to a free choice procedure, when a gustatory aversive marker is introduced, the preference is guided more by taste than by any learned appreciation of pharmacological effects

## Experiment 5: Taste Relevance

To check the relevance of the gustatory component and verify whether animals were unable to choose simply because the two purposely sweetened solutions had the same taste, a gustatory aversive marker was introduced. Rats were submitted to the same free-choice procedure described in Experiment 1, but in this case quinine sulfate (Q) (100 μg/ml) was added to the bottle containing E for one group of animals or to Asp for the other. Fluid intake, analgesia and catatonia were evaluated as described in experiment 1. Classical morphine was also offered for 24 hours, as described in the Experiment 3, using a crossed experiment. During the first part of the experiment animals were offered morphine HCl (M) (500 μg/ml) as an alternative to pure Asp and, during the second part, M was offered as an alternative to a Q-Asp solution.

## Results

When the marker was added to the medicated solution, rats, allowed to drink for 1 h, showed a preference for Asp over a period of 45 days, even though the small amount of ingested E led to analgesia with tolerance. When a withdrawal syndrome was elicited (total score=11), and the same choice was offered 1 hour later, the preference was unchanged. On the contrary, when the aversive marker was added to Asp solution, the animals showed a significant preference for the E solution. In this case, the large volume ingested of E led to such a high degree of analgesia associated with catatonia that no tolerance was elicited, and all the rats were put to death for respiratory depression (data not shown). Also using a schedule of 24 h and the classical M (Fig.4), the preference was guided more by taste than by any learned appreciation of pharmacological effects. In fact, during the first 23 days, when the M-Asp solution was offered as an alternative to pure Asp, the M solution was refused; however, when it was offered as an alternative to a Q-Asp solution, the M solution was preferred.

## General Discussion

The general conclusion which can be drawn from our findings is that rats seem unable to discriminate between the offered fluids with the same taste. In fact, the amounts of drugs of abuse taken daily by oral route in a given free-choice setting: 1) remain more or less constant for days; 2) are unrelated to neurodepressant or stimulant effects; 3) they are not affected by premedication and do not change in presence of either tolerance or sensitization; do not increase when the withdrawal is elicited. These unexpected findings are supported by several reports in which only some monkeys offered morphine dissolved in drinking water, chose the opiate (3, 11). More recently, Ronnback (23), reporting individual differences in morphine preference within the Sprague-Dawley strain, has not found significant differences in morphine serum levels, taste sensitivity, severity of abstinence reactions in comparison to low-morphine preference rats. Ator and Griffiths (24), reported that premedicated rats with barbiturate or benzodiazepine solutions did not choose them if they were concurrently offered plain water. Furthermore, Falk et al.(25) have shown that cocaine is not chosen in preference to water when a polydipsic schedule is used. Finally, George (26) found no significant positive relationship between ethanol preference and ethanol reinforcement in different strains of alcohol preferring rats. The surprising conclusion which we can draw, is that rats failed to associate these central effects with what they have drunk even after many days. However, the amounts taken may be dramatically modified by a change in the taste of the solutions. In other words, the perception of peripheral factors seems to outweigh the perception of central effects. In a free choice setting, a partial preference for the solutions containing drugs of abuse may be obtained by prolonging the time of assumption or by allowing intermittent assumption. However, this preference also seems to be unrelated to the intensity or duration of elicited central effects, suggesting that the time of contact and the contingency of the stimulus can help animals in discriminating medicated solution devoid of aversive taste.

### References


1. Siegel R K. Intoxication: Life in pursuit of artificial paradise, EP Dutton, New York, 1989.
2. Brady J V, Fischman M W. Assessment of drugs for dependence potential and abuse liability: an overview. In: Behavioral pharmacology: the current status, AR Liss Inc., New York, 1985, pp 361-382..
3. Deneau G, Yanagita T, Seevers M H. Self-administration of psychoactive substances by the monkey. Psychopharmacologia 16:30-48, 1969.
4. Goldberg S R, Henningfield J E. Reinforcing effects of nicotine in humans and experimental animals responding under intermittent schedules of IV drug injection. Pharmacol Biochem Behav 30:227-234, 1988.
5. Falk J L. Production of polydipsia in normal rats by intermittent food schedule. Science

133:195-196, 1961.
6. Leander J D, McMillan D E, Harris L S. Schedule-induced oral narcotic self-administration acute and chronic effects. J Pharmacol Exp Ther 195:279-287, 1975.
7. Tang M, Falk J L. Ethanol dependence as a determinant of fluid preference. Pharmacol Biochem Behav 7:471-474, 1977.
8. Tang M, Falk J L. Oral self-administration of cocaine: chronic excessive intake by schedule induction. Pharmacol Biochem Behav 28:517-519, 1987.
9. Falk J L, Tang M J. Schedule induction of drug intake: differential responsiveness to agents with abuse potential. Pharmacol Exp Ther 249:143-148, 1989.
10. Maradones J R, Segovia-Riquelme N, Hederra A D, Alcaino F G. Effect of some self-selection conditions on the voluntary alcohol intake of rats. Quart J Stud Alc 16:425-437, 1955.
11. Claghorn J L, Ordy J M, Nagy A. A spontaneous opiate addiction in rhesus monkeys. Science 149:440-442, 1965.
12. Kumar R, Steinberg H, Stolerman I P. How rats can become dependent on morphine in the course of relieving another need. In: Scientific basis of drug dependence, edited by Steinberg H. Churchill Ltd., 1969, pp 209-220.
13. Khavary K A, Peters T C, Baity P L. Voluntary morphine ingestion, morphine dependence, and recovery from withdrawal signs. Pharmacol Biochem Behav 3:1093-1096, 1975.
14. Chernov H I, Barbaz B S, Bosher R L, Feist M N. Age and lack of handling as factors in the consumption of an etonitazene solution by naive rats. Arch Int Pharmacodyn 195:231-239, 1972.
15. Sala M, Braida D, Calcaterra P, Leone M P, Gori E. Possibility of spontaneous drug abuse tested in rat. Pharmacol Res 28(1):21-34, 1993.
16. Myers R D. Alcohol choice in Wistar and G-4 rats as a function of environmental temperature and alcohol concentration. J Comp Physiol Psychol 54:510-516, 1961.
17. Sala M, Biagetti R, Braida D, Maggioni A, Gori E. La polidipsia da condizionamento aggiuntivo o da angiotensina centralmente somministrata quale modello di dipendenza da oppiacei. In: Atti 1° Congresso CNR, Sottoprogetto Tossicodipendenze, 1985, pp 141-150.
18. D'Amour F E, Smith D L. A method for determining loss of pain sensation. J Pharmacol Exp Ther 72:74-79, 1941.
19. Dunstan R, Broekkamp C L, Lloid K G. Involvement of the caudatus nucleus, amygdala or reticular formation in neuroleptic and narcotic catalepsy. Pharmacol Biochem Behav 14:169-174, 1981.
20. Huang J T, Yano I, Takemori A E. Effect of various central biogenic amine-modifiers and ambient temperature on the naloxone-induced jumping in morphine-dependent mice. In: Factors affecting the action of narcotics, edited by Adler M R, Manara L, Samanin R. Raven Press, New York, 1978, pp 495-509.
21. Wikler A, Martin W R, Pescor F T, Eades C G. Factors regulating oral consumption of an opioid etonitazene by morphine-addicted rats. Psychopharmacologia 5:55-76, 1963.
22. Grant S, Redmond D E. Clonidine suppresses methylxanthine induced quasi-morphine withdrawal syndrome. Pharmacol Biochem Behav 17:655-658, 1982.
23. Ronnback L. Morphine preference in individual rats after morphine ingestion. Psychopharmacology 102:257-262, 1990.
24. Ator N A, Griffiths R R. Self-administration of barbiturates and benzodiazepines: a review. Pharmacol Biochem Behav 27:391-398, 1994.
25. Falk J L, Vigorito M, Tang M, Lau C E. Schedule-induced cocaine drinking: choice between cocaine and vehicle. Pharmacol Biochem Behav 35:187-193, 1990.
26. George F R. Genetic approaches to studying drug abuse: correlates of drug self-administration. Alcohol 7:207-211, 1990.

# 6
# Methadone Maintenance Comes of Age

**V. P. Dole**
The Rockefeller University, New York, NY, U.S.A.

**Summary**

Methadone maintenance treatment in the United States is reaching maturity. During the past three decades it has progressed from an innocent childhood, through a turbulent adolescence, to recognition as an essential medical procedure. At present approximately 115.000 former heroin addicts in the United States are being treated in 750 clinics located in 40 of the 50 states. The treatment has survived challenge by professional sceptics, by ideologically hostile agencies, by competitive modalities, and even by well-intentioned clinicians in methadone programs who have prescribed inadequate doses of the medication.

The simple fact is that it works (1-8). When given to a heroin addict once daily in adequate dose, methadone abolishes the compulsive narcotic hunger of addicts without producing euphoria. The transformation of apparently hopeless narcotic users into normally functioning individuals seemed miraculous to the original research team -- Marie Nyswander, Mary Jeanne Kreek and me (9) -- three decades ago. How could the administration of a narcotic medicine normalize persons whose dependence on narcotics had been characterized by psychiatrists as immature, degenerate, sociopathic, etc?

The phenomenon obviously demanded verification and further study under different conditions to determine its potential as a large-scale treatment. A talented group of clinicians, nurses, administrators, ex-addicts, vocational specialists, epidemiologists and lawyers joined us in studying the question. The list of persons who participated in the work during the first decade is too long for proper acknowledgment here, but the critically important involvement of persons closest to the early work must be mentioned: Drs. Joyce Lowinson, Ray Trussell, Robert Newman, Lars Gunne, Robert Millman, Elizabeth Khuri (physicians); Maurice Bachrach (administrator); Leo Vono (vocational specialist); Frances Gearing, Herman Joseph (epidemiologists); Donal O'Brien, Manny Guerrero (lawyers); and a team of ex-addicts who must remain anonymous. Systematic studies were initiated with financial support from New York City's Health Research Council and Department of Hospitals with endorsement by civic leaders.

The work moved forward rapidly. Over the next ten years treatment programs utilizing methadone were established in Beth Israel Medical Center and Bronx State Hospital with other major hospitals of New York City serving as administrative and medical backup centers supporting a network of satellite clinics that served all of the city's boroughs. Physical examinations on admission and periodically thereafter defined the medical needs of all patients, and were established as a routine to detect any ill effects of the medication. None were found (10,11). A comprehensive, computerized data system was developed to record the progress and problems of every person admitted, with repeated analyses of urine for drugs to provide objective data on response to treatment. An independently appointed evaluation committee was empowered to inspect records, interview patients and publish independent reports, subject only to the requirement that the names of the patients be kept confidential. Professionals in the field of drug abuse were welcomed as visitors to the clinics, and a series of National Conferences on Methadone Treatment was inaugurated in 1968 to provide a forum for reporting experience in other cities.

The first decade culminated in the admission (during 1970-74) of over 20,000 heroin addicts into a diversified system of clinics in many hospitals of New York City, directed by Robert Newman and supported by the city's Health Services Administration (12). The public health impact of this major program was reflected in statistics showing a significant reduction in arrests for drug related crimes and a reduction in reported cases of hepatitis (presumably due to a decrease in intravenous drug abuse).

Thus by the end of the first decade after the small experimental beginnings in a metabolic ward a successful model for large-scale treatment of heroin addicts was in place. Maintenance treatment had been proven to be medically safe, feasible to administer on a large scale and generally successful in rehabilitation of addicts. Needless to say, throughout this period the treatment had been under continuous attack by supporters of the old regime -- the lockup and punish school of therapists -- and by bureaucrats whose authority was threatened by the new program. Complicating matters, the State of New York at the same time was funding a hugely expensive and unsuccessful lockup program that it had been advertised in political campaigns as a definitive answer to the addiction problem.

Opposition from vested interests did not block progress in developing the methadone program (in fact it contributed to the unity of the staff), but we were surprised by the indifference of the medical profession and the failure of public health officials to provide strong support for expansion of a treatment that had proven successful. In the second decade, instead of expansion, the objectives of treatment were reversed by Federal agencies and the authority of physicians was limited by these agencies on political and philosophical grounds: Abstinence rather than rehabilitation was restored as the goal of treatment; doses were lowered to levels that were frequently inadequate; administrators became punitive and often contemptuous of the patients (now called "clients"); termination of maintenance was encouraged despite an 80% relapse rate. The punitive attitude of moralists hardened into regulatory law, subjecting clinics to an unprecedented degree of control over every detail of their operation. In New York, in 1976 at the peak of this folly, a survey revealed that methadone clinics were subject to regulation by 11 uncoordinated agencies of Federal, State, City and Community, and that a physician working in the clinic would need to spend 185% of his time in paperwork to provide the redundant reports and documentation required by these different agencies (13).

The quality of treatment deteriorated during this period (14). Underfunded, crowded, operating in poor quarters, prevented from moving to more adequate facilities by hostile communities, harassed by teams of inspectors who criticized their deficiencies without providing money or political support for improvement, with a negative image fueled by disinformation in the media, the methadone clinics nevertheless survived, thanks to the dedication of their overworked staffs. While under this pressure methadone clinics were providing more than 90% of all the long-term rehabilitation services available to heroin addicts in New York City.

This turbulent phase has not ended, but the third decade brought a change in political attitude (15). The onset of the AIDS epidemic in the late 1970s alerted public health officials to the fact that needle-sharing addicts are a major vector in transmission of the deadly virus (as well as other blood-borne infections), and that AIDS patients, with impaired immunity, constitute a large susceptible population into which drug-resistant tuberculosis is spreading.

Even the opponents of maintenance treatment have been obliged to concede that no alternative treatment has sufficient capacity and proven efficacy to play a significant role in curbing transmission of the virus by addicts. Today there is a new generation of clinicians with knowledge of recent developments in neurophysiology. Younger physicians with more training in neurophysiology are prepared to recognize drug craving as a symptom of a neurochemical imbalance, and therefore to accept maintenance treatment as rational chemotherapy. The field of addiction medicine has emerged from the rigid dogma of descriptive psychiatry ("character defect", "sociopathic personality", etc) into the modern world of neurophysiology. Methadone maintenance is part of this new field -- indeed it might qualify as one of the founding members -- but its role is transitional. In time, with enough effort and luck, the neurochemical disorders responsible for the pathological craving will be identified, and other medications for treatment of addictions will be developed that are simpler to use, and perhaps even curative.

We enter the fourth decade of maintenance treatment with new and difficult challenges but with more knowledge (16). The explosive spread of AIDS --a disease so far without effective immunization or treatment -- has amplified the problem of drug abuse, and made effective treatment of addicts a critical priority. Needle-using addicts have become a threat to the health of the general population, first spreading infection among themselves and subsequently to others by sexual activity and contamination of the blood supply. In effect, addiction makes AIDS a highly contagious disease.

The challenge for the immediate future is to make effective use of what we have learned about addictions in the past few decades. Of course it would be easier if we had a vaccine or other medication that would eliminate the susceptibility of all potential addicts -- not an impossible hope since only a minority of person become chronically addicted to narcotics or alcohol when exposed -- and a definitive cure for established addiction is a dream worth pursuing, but as a practical matter we are confronted at present with serious public health problems requiring immediate responses. A person with responsibility for health of a community must work with the tools that are available.

They are considerable. If the lessons of the past are learned, large-scale programs providing adequate doses of methadone, and social support as needed can rehabilitate heroin addicts. They do not need to be coercive since chronic heroin addiction is not pleasant, but they must be readily available and have outreach to alienated, homeless addicts. Although methadone has only a specific and limited effect, the therapeutic environment of a good clinic provides a favorable setting for treatment of complicating addictions and infections.

Termination of needle use by addicts reduces the transmission of AIDS. Social rehabilitation of addicts reduces crime and improves neighborhoods, thus reducing the availability of addictive drugs in the area. Rehabilitated ex-addicts, having escaped the slavery of heroin addiction, can become highly motivated and effective agents of prevention in a community susceptible adolescents.

This sounds straightforward, but the practical difficulties of developing large-scale treatment facilities should not be minimized. To be effective as a public health response, maintenance clinics must have a capacity that is comparable to the size of the addict population and yet continue to be as effective in rehabilitation as they were when small. Expansion of services will face prejudice, denial, hostility and indifference -- but with informed medical leadership the nature of the addiction problem can be explained to politicians and the public. The process of developing an effective response to the many problems of addiction starts with discussion among specialists in meetings such as this, then expands to inform the general medical profession and the public.

Now, the time has arrived in this meeting for the real experts to come forward and lead the discussion -- the clinicians who are treating patients today. I am reminded of a parallel situation in 1955 when the co-founder of Alcoholics Anonymous, Bill W., spoke on the occasion of the twentieth anniversary of the Fellowship and saluted the "Coming of Age of AA" (17). Like Bill on that occasion, I look to the future of the Methadone Maintenance Program with confidence, knowing that it too, has Come of Age.

## References


1. Dole V P. In the course of professional practice. N Y State J Med 65:927-931, 1965.
2. Dole V P, Nyswander M E. A medical treatment for diacetylmorphine (heroin) addiction. J Am Med Assoc 193:646-650, 1965.
3. Dole V P, Nyswander M E, Warner A. Successful treatment of 750 criminal addicts. J Am Med Assoc 206:2708-2711, 1968.
4. Dole V P. Narcotic addiction, physical dependence and relapse. New Engl J Med 286:988-992, 1972.
5. Dole V P. Methadone maintenance treatment for 25.000 heroin addicts. J Am Med Assoc 215:1131-1134, 1971.
6. Dole V P. Detoxification of sick addicts in prison. J Am Med Assoc 220:366-369, 1972.
7. Dole V P. Addictive behavior and the art of medicine, The 14th Leonard Ball Oration, Australia, 1982.
8. Ball J C, Ross A. The Effectiveness of Methadone Maintenance Treatment, Springer-Verlag, New York, 1991.
9. Dole V P, Nyswander M E, Kreek M J. Narcotic blockade. Arch Intern Med 118:3004-3009, 1966.
10. Dole V P. Methadone maintenance. A report of two years of experience. In: Committee on Problems Drug Dependence, NAS NRC, 1966,

11. Dole V P, Nyswander M E. Behavioral pharmacology and treatment of human drug abuse: Methadone maintenance of narcotic addicts. In: The Neurobiology of Opiate Reward Processes, edited by Smith J E, Lane J D. Elsevier Biomedical Press, Amsterdam, 1983, pp 211-233.
12. Newman R G. Methadone Treatment in Narcotic Addiction. Program Management, Findings and Prospects for the Future, Academic Press Inc., New York, 1977.
13. Dole V P, Vogelson P, Leonigsberg L, Bachrach M. A survey of paperwork, The Community Treatment Foundation (Unpublished), 1976.
14. Dole V P, Nyswander M E. Methadone maintenance treatment. A ten-year perspective. J Am Med Assoc 235:2117-2119, 1976.
15. Dole V P. Implications of methadone maintenance for theories of narcotic addiction. The Albert Lasker Medical Awards. J Am Med Assoc 260:3025-3029, 1988.
16. Dole V P. Addiction as a public health problem (Presented to the American Society for Addiction Medicine, April 1991). Alcoholism: Clin Exp Res 15:749-752, 1991.
17. Dole V P. What have we learned from three decades of methadone maintenance treatment. (In press). Drug and Alcohol Review 13:1994.
18. Alcoholics Anonymous. Comes of Age, Alcoholics Anonymous Publishing, Inc., New York, 1957.

# 7

# Pharmacotherapy for Craving

**I. Maremmani[1], O. Zolesi[1,2], L. Daini[1,2], M.R. Capone[1], M. Aglietti,[1] P. Castrogiovanni[1,2]**
[1]Dependence-Psychiatry Unit, Psychiatric Institute, University of Pisa, Italy
[2]Ph.D. Research Program Siena, Cagliari & Pisa Universities, Italy

## Summary

The term craving derives from common language and its use in scientific literature is still under discussion. Nevertheless it seems appropriate for describing the very essence of "addiction" as compulsion. The various clinical forms of craving have two main features: a potent reinforcing effect which is at the base of drug addiction and a negative reinforcement mechanism linked to phobic feelings aimed at avoiding withdrawal symptoms. Anticraving therapy should act on both of these aspects. After discussing the results obtained with the principal anticraving drugs, such as methadone, GHB, dopamine agonists (bromocriptine), desimipramine, mood stabilizers, and serotoninergic drugs, the results of personal studies in progress are discussed.

At present, knowledge about drug dependence seems to be characterized by a common denominator, craving. Craving could thus be defined a new psychopathological or syndromic entity which Marlatt argues, captures the essence of addiction, in terms of its compulsive qualities (1).

Wise identifies 2 types of craving on the basis of reinforcement. One type of craving is associated with positive reinforcement and is considered fundamental for the initiation of drug use, for the transition from use to abuse and may be responsible for relapse after a period of abstinence. The other type of craving is associated with a phobic mechanism, (which is a negative reinforcement) that aims at avoiding withdrawal symptoms, and is an important factor in maintaining dependence to opioids in particular (2,3).

The phenomena of craving has the following characteristics:

1. Strong desire for a substance or a situation.
2. Somatic symptoms related to the autonomic nervous system.
3. Drug or situation seeking behavior
4. Inability to stop drug or situation seeking behavior in spite of obstacles, physical danger or social consequences.

If craving is the base of addiction in the beginning of the natural history of drug use as well as in continued drug use ( increasing doses and revolving door phase) its treatment may be essential to the "management" of drug dependence.

Observations that various substances have a common biological pathway of action related to craving opens the way to interesting hypotheses regarding polydrug abuse and the possibility of more reasonable therapeutic interventions. In fact, different substances stimulate the same neurotransmitter systems, triggering a positive reinforcement (4-12,2). The physical needs and craving experienced by the drug addict could be partially satisfied by other substances thus explaining polydrug abuse. The practice of polydrug abuse may arise not only because the substance of choice is not available but because craving has been inadequately treated.

The pharmacological agent suitable for the treatment of craving should alleviate withdrawal symptoms and interact with the positive reinforcement system. Most pharmacological agents used in the treatment of craving do not have these characteristics. For example clonidine is effective on withdrawal symptoms in heroin drug addicts, reducing the phobic component of craving by interacting with the Locus Coeruleus but has no effect on the pathways involved with positive reinforcement (13,14).

**Methadone**

The cornerstone in heroin dependence treatment is long term medication with opioid agonists, methadone, which have the greatest anticraving effect and create tolerance to narcotic action. Methadone's anticraving effect is not only due to its action on withdrawal symptoms nor to blocking the pharmacological effects of heroin, but more specifically to its action on the positive reinforcement mechanism.

Antagonist drugs block the euphoric effects of heroin, prevent tolerance and thus withdrawal yet addicts treated with antagonists have a kind of craving that manifests as dysphoretic mood and discomfort which frequently leads to interrupting treatment and relapsing to heroin use.

There are 3 possible treatment dosages with Methadone:

1. Withdrawal syndrome dosage (30mg/daily) which does not block opioid receptors and does not prevent relapses.
2. Blocking dosage. Euphoria is not experienced with heroin (60-80mg), although in this phase the subject may continue drug use. Drug seeking behavior may persist and the subject may use BDZ or alcohol to feel the effects of heroin.
3. Anticraving dosage (80-120 mg/daily) blocks craving which is at the base of drug seeking behavior (15).

The effective dosage capable of blocking opioid receptors is relatively constant whereas the anti-craving dose is more subjective. There is no pre-established anti-craving limit and clinical evidence reports of subjects who experienced this limit at much higher dosages. We had one case of heroin drug addiction treated with three hundred and eighty mg of methadone daily.

Concurrent psychic or social factors may also contribute. Craving is observed when drug free subjects talk about heroin in group therapy or when they returned to areas where they had previously used drugs (so called reflex abstinence). It is also experienced by subjects in methadone maintenance when they experience losses or life changes. In this case craving is experienced as a discomfort which requires a higher methadone dose for a short period of time.

**GHB (Gamma Hydrossibutirric Acid)**

GHB was proposed as anti-craving treatment in alcoholism; it was more effective than placebo in a double blind study (16). Fadda reported that GHB suppressed voluntary intake of alcohol in selected drinking rats and prevented withdrawal symptoms in rats who were alcohol dependent. Recent research suggests that GHB reduces the desire for alcohol by reproducing its pleasurable effects in a gradual and stable way (17,18). GHB interacts with various neurotransmitters. Tunnicliff maintains thatGHB derives from Gaba (19). Other authors exclude this possibility (20). According to Tunnicliff GHB may modulate dopaminergic activity in the nigra substance (19). Mamelak has found an increase in cerebral DA in experimental animals after administration of GHB (21). At low doses GHB appears to stimulate electrical activity of dopaminergic neurons (22). Some researchers are not in agreement with this observation suggesting the use of GHB in the treatment of schizophrenia given its inhibiting action on the dopaminergic system. Experimental data report that Narcan eliminates the effects produced by GHB on dopamine, observed in EEG and on behavior; this suggests involvement of the endogenous opioid system in the biological activity of GHB (23).

In our studies we found that one heroin addict in treatment with antagonists and antidepressants who relapsed to heroin use (the patient was not an alcohol abuser) treated with 20 cc/daily of GHB stopped heroin use. The anticraving effect is well documented in a detoxified heroin drug addict who controlled his craving with high dosages of BDZ. When treated with 80 cc of GHB and 30 mg of methadone daily he stopped using BDZ. The anticraving effect of GHB was also noted in an undermedicated (30 mg.) addict who also abused alcohol; we observed 60% reduction in alcohol intake after GHB (30 cc/daily) was included in his therapy. Finally GHB was shown to be effective in 20 depressed alcoholics. Previously treated with antidepressants, mood disorder had improved but not alcohol intake. When GHB (average = 25 cc) was added alcohol intake was reduced by about 65%(24).

**Dopamine Agonists**

The use of dopamine agonists in the treatment of drug addiction finds support in the relationship hypothesized between craving and the dopaminergic system. Bromocriptine is indicated as a possible therapeutic approach in cocaine dependence (25-30). But the adverse effects of this medication limit its use . An alternative to bromocriptine is pergolide with effective anticraving action and good results have also been obtained with desimipramine (31).

## Mood Stabilizers

The hypotheses that kindling induced by cocaine is the biological base of craving introduces anti epileptic drugs such as carbamazepine and valproic acid into the realm of "craving" therapy. Halikas has obtained encouraging results with valproic acid in the treatment of cocaine dependence (32). Good results have also been reported for carbamazepine (33) . With regards to Lithium, some researchers report that it behaves as an opioid antagonist preventing or modifying euphoric mood although unable to block euphoric effect produced by morphine, as Narcan does (34,35). Other authors argue that long term effects of lithium could potentiate opioid agonists (36-38). Doses of 1.2 m.Eq. block cocaine induced euphoria and have an anticraving effect (39,40,33). Buprenorphine, an opioid agonist-antagonist, has an anticraving effect for cocaine but it is unclear if it reduces craving for opioids as methadone does (41).

Serotoninergic Drugs

The observation that craving may be strictly related to disregulation of the serotoninergic system has encouraged research in the use of serotonin reuptake inhibitors. Fluoxetine seems to be effective in alcoholism, in cocainism, in bulimia (42-46). We have recently observed the prospect of fluoxetine's anticraving action in heroin dependence (47). 13 heroin addicts treated with naltrexone and fluoxetine showed a greater probability of 6 months survival time (no relapses) when compared to 12 addicts treated only with antagonists (personal unpublished data).

## Conclusions

Although clinical and therapeutic aspects of craving are by no means fully understood, one thing appears certain. The physiopathological mechanisms at the base of this phenomena and the cross tolerance among the various substances, permits a better understanding of otherwise incomprehensible clinical situations, i.e. polydrug abuse, or more specific situations (cocaine craving in heroin addicts, or carbohydrates craving in ex smokers) and permits a better understanding of the repercussions on prognosis and therapy.

## References


1. Marlatt A. Craving notes. Br J Addict 82:42-44, 1987.
2. Wikler A. Opioid Dependence, Plenum, New York, NY, 1980.
3. Shaikh M B, Brutus M, Siegel H E, Siegel A. Differential control of aggression by the mid brain. Exp Neurol 83:436-442, 1984.
4. Koob G F, Bloom F E. Cellular and molecular mechanism of drug dependence. Science 242:715-723, 1988.
5. Risner M E, Jones B E. Role of noradrenergic and dopaminergic processes in amphetamine self-administration. Pharmacol Biochem Behav 5:477-482, 1976.
6. Koob G F. The dopamine anhedonia hypothesis: a pharmacological phrenology. Behav Brain Sci 5:63-64, 1982.
7. Risner M E, Jones B E. Intravenous self-administration of cocaine and norcocaine by dogs.

Psychopharmacology 71:83-89, 1980.

8. Bozart M A, Wise R A. Involvement of the ventral tegmental dopamine system in opioid and psychomotor stimulant reinforcement. In: Problems of drug dependence, edited by Harris L S. N.I.D.A., Washington, DC, 1985, pp 190-196.
9. Zito K A, Vickers G, Roberts D C S. Disruption of cocaine and heroin self administration following kainic acid lesions of the nucleus accumbens. Pharmacol Biochem Behav 25:1029-1036, 1985.
10. Philiphs A G, Broekkamp C L E, Fibiger H C. Strategies for studying the neurochimical substrates of drug reinforcement in rodents. Prog Neuro-psychopharmacol & Biol Psychiat 7:585-590, 1983.
11. Ettenberg A, Camp C H. Haloperidol induces a partial reinforcement extinction effects in rats: implication for a dopamine involvement in food reward. Pharmacol Biochem Behav 25:813-821, 1986.
12. Ettenberg A, Camp C H. A partial reinforcement extinction effect in water-reinforced rats intermittently treated with haloperidol. Pharmacol Biochem Behav 25:1231-1235, 1986.
13. Aghajanian G K. Tolerance of Locus Coeruleus neuron to morphine and suppression of withdrawal response by clonidine. Nature 276:186-188, 1978.
14. Gold M S, Kleber H D. Clinical utility of clonidine in opiate withdrawal; a study of 100 patients. In: Pharmacology of clonidine, edited by Harbans L, Fielding S. Alan Liss, New York, NY, 1981, pp 299-306.
15. Maremmani I, Castrogiovanni P. Manuale del Trattamento Ambulatoriale con Metadone, SIMS, CN-MPC, Pisa, 1992.
16. Gallimberti L, Ferri M, Ferrara S D, Fadda F, Gessa G L. Gamma-hydroxybutyric acid in the treatment of alcohol dependence: A double-blind study. Alcohol Clin Exp Res 16:673-676, 1992.
17. Fadda F, Argiolas A, Melis M R, et al. . Differential effect of acute and chronic ethanol on dopamine metabolism in frontal cortex, caudate nucleus and substantia nigra. Life Sci 27:979-986, 1980.
18. Gessa G L. Nascita di un farmaco per l'alcoolismo: l'acido gamma-idrossibutirrico. In: Il GHB nel trattamento della dipendenza alcoolica, edited by Gallimberti L, Ferrara S D, Gessa G L. Addiction Research Foundation of Italy, Padova, 1992, pp 3-4.
19. Tunnicliff G. Significance of gamma-hydroxybutyric acid in the brain. Gen Pharmacol 23:1028-1034, 1992.
20. Serra M, Sanna E, Foddi C, Concas A, Biggio G. Failure of gamma-hydroxybutyrate to alter the function of the GABA(A) receptor complex in the rat cerebral cortex. Psychopharmacology 104:351-355, 1991.
21. Mamelak M. Gamma-hydroxybutyrate: an endogenous regulator of energy metabolism. Neurosci Biobehav Rev 13:187-198, 1989.
22. Ferrara S D, Gessa G L, Gallimberti L. Farmacotossicologia e farmacocinetica del GHB. In: Il GHB nel trattamento della dipendenza alcoolica, edited by Gallimberti L, Ferrara S D, Gessa G L. Addiction Research Foundation of Italy, Padova, 1992, pp 5-14.
23. Crosby G, Ito M, Kaufman E, et al. . Naloxone pretreatment alters the local cerebral metabolic effect of gamma-hydroxybutyrate in rats. Brain Res 275:194-197, 1983.
24. Zolesi O, Daini L, Capone M R, Aglietti M, Nardini C F, Maremmani I, Castrogiovanni P. The use of GHB in anticraving therapy. Preliminary data in alcoholic depressed patients, in alcohol and polydrug use of heroin addicts during MMTPs and in heroin craving during opioid antagonist therapy, Second European Symposium on Drug Addiction & AIDS, Siena, 1993.
25. Malcolm R, Hutto B R, Phillips J D, Ballenger J C. Pergolide mesylate treatment of cocaine withdrawal. J Clin Psychiatry 52:39-40, 1991.

26. Extein L, Allen S, Gold M S, Morales A, Goodnick P J, Gross D A. Comparison of bromocriptine and desimipramine in cocaine withdrawal. Ann Clin Psychiatry 1:193-197, 1989.
27. Kumor K, Sherer M, Jaffe J. Effects of bromocriptine pretreatment on subjective and physiological responses to IV cocaine. Pharmacol Biochem Behav 33:829-837, 1989.
28. Teller D W, Devenyi P. Bromocriptine in cocaine withdrawal - Does it work? Int J Addict 23:1197-1205, 1988.
29. Dackis C A, Gold M S, Sweeney D R, et al.. Single dose bromocriptine reverses cocaine craving. Psichiat Res 20:261-264, 1987.
30. Berg G, Laberg J C, Skutle A, Ohman A. Instructed versus pharmacological effects of alcohol in alcoholics and social drinkers. Behav Res Ther 19:55-66, 1981.
31. Gawin F, Kleber H. Pharmacological treatments of cocaine abuse. Psychiatr Clin North Am 9:573-583, 1986.
32. Halikas J A, Crosby R D, Carlson G. Valproate in the treatment of cocaine addiction: A preliminary report. Ann Clin Psychiatry 4:65-66, 1992.
33. Kosten T R, Gavin F, Rounsaville B J, Kleber H D. Cocaine abuse among opioid addicts. Amer J Drug Alcohol Abuse 1987.
34. Jasinski D R, Nutti J G, Haertzen C A, Griffith J D. Lithium: Effects on subjective functioning and morphine-induced euphoria. Science 195:582-584, 1977.
35. Snyder S H, Simantox R. The opiate receptor and opiate peptides. J Neurochem 28:13-20, 1977.
36. Jensen J. The effect of prolonged lithium ingestion on morphine actions in the rat. Acta Pharmacol Toxicol 35:395-402, 1974.
37. Judd L L, Hubbard B, Janowsky D S, et al.. The effect of lithium carbonate on affect, mood and personality of normal subjects. Arch Gen Psychiatry 34:346-351, 1977.
38. Mirin S R, Meyer R E, McNamee B. Psychopathology and mood duration in heroin use: Acute and chronic effects. Arch Gen Psychiatry 33:1503-1508, 1976.
39. Gawin F H, Ellinwood E H. Cocaine dependence. Annu Rev Int Med 40:149-162, 1989.
40. Gawin F H, Kleber H D. Cocaine abuse treatment: open pilot trial with desimipramine and lithium carbonate. Arch Gen Psychiatry 41:903-909, 1984.
41. Mello N K, Mendelson J H, Kuehnle J C, Sellers M S. Operant analysis of human heroine self-administration and the effects of naltrexone. J Pharmacol Exp Ther 216:45-54, 1981.
42. Gorelick D A, Paredes A. Effect of fluoxetine on alcohol consumption in male alcoholics. Alcohol Clin Exp Res 16:261-265, 1992.
43. Pollack M H, Rosenbaum J F. Fluoxetine treatment of cocaine abuse in heroin addicts. J Clin Psychiatry 52:31-33, 1991.
44. Wurtman J J, Wurtman J R, Mark S, et al. . D-fenfluramine selectively suppresses carbohydrate snacking by obese subjects. Int J Eating Disord 4:89-99, 1985.
45. Wurtman J J, Wurtman J R, Reynolds S, et al. . D-fenfluramine suppresses snack intake among carbohydrate cravers but not among non carbohydrate cravers. Int J Eating Disord 6:687-699, 1987.
46. Levine L R, Pope H G J, Enas G G, Wilson M G, Ballenger J C, Blinder B J, Dixon K, Dunlop S R, Ferguson J M, Goldbloom D S, Halikas J, Herzog D B, Hsu L K G, Hudson J I, Powers P S, Reimherr F W, Yates W R. Fluoxetine in the treatment of bulimia nervosa: A multicenter, placebo- controlled, double-blind trial. Arch Gen Psychiatry 49:139-147, 1992.
47. Maremmani I, Daini L, Zolesi O, Castrogiovanni P. Use of Fluoxetine in heroin addiction. Br J Psychiatry 160:570-571, 1992.

# 8
# The Use of GHB in Anticraving Therapy. Preliminary Experimental Data of its Use in Alcohol and Heroin Dependence

**O. Zolesi[1,2], L. Daini[1,2], M. R. Capone[1], M. Aglietti[1], F. Raimondi[1], I. Maremmani[1]**
[1]Dependence-Psychiatry Unit , Psychiatric Institute, University of Pisa, Italy
[2]Ph.D. Research Program Siena, Pisa & Cagliari Universities

## Summary

GHB seems to be an effective anti-craving agent not only in alcoholic patients, as already demonstrated, but also in other drug addiction pathologies. Open studies, carried out at the Dependence-Psychiatry Unit of the Psychiatric Clinic of University of Pisa, showed its effectiveness and lack of side effects when associated with anti-depressants and mood stabilizers, in 20 depressed alcoholic subjects resistant to anti-depressant therapy. The anti-craving effect was also observed in 2 heroin drug addicts, who were abusing alcohol and benzodiazepines during methadone maintenance treatment, and in 4 subjects treated with naltrexone, who were still "craving" heroin. The average observation period was 31 days, the average dosage was 35 ml/die, min 15 max 60; no significant side effects were observed.

## Introduction

Preliminary data are presented of an open study still in progress at the Day Hospital of the Dependence-Psychiatry Unit of the University of Pisa evaluating the efficacy and tolerability of GHB as anticraving therapy in chronic alcoholism and heroin dependence.

A naturalistic approach was used in this open study. Severity of illness, global improvement and efficacy index were evaluated with the Clinical Global Impressions (CGI). Pharmacological agents, doses, modification of treatment plan, side effects and variations of substance intake were registered with the Schedule for Patient Monitoring (SPM). Patients were evaluated at baseline and each time that therapy was modified with the CGI and SPM.

### A. Effectiveness and Tolerability in 20 Subjects with Double Diagnosis of Alcoholism and Major Depression, Recurrent

Subjects with double diagnosis of alcoholism and depression were considered. Those who were included in the study met DSM-III-R criteria for Major Depression, Recurrent and Alcohol Dependence. They had previously been treated with antidepressants and/or mood stabilizers for an adequate period of time at therapeutic doses. Their affective disorder had responded favorably to treatment but alcohol intake remained unmodified Criteria of exclusion were: serious organic pathologies, psychoses or melancholic states.

The data presented refer to a sample of 20 subjects, 10 males and 10 females, mean age 46 years (sd 10; min 28, max 65). Mean length of observation period at endpoint was 72 days (sd 53; min 8 max 175)

The severity of illness at baseline ranged between 4 and 7 (XM=5 sd=1.1) and thus these subjects were moderately to most extremely ill. Alcohol intake was considered by default 100%. However the minimum daily intake considered was 2cc/Kg. Depressive symptomatology ranged from moderate to severe. 16 psychoactive agents were associated with GHB. No patients was treated only with GHB.

**Table 1.** Paired T-Test (1st and last assessment) in 20 depressed alcoholics treated with GHB and other psychoactive agents)

| | Baseline | | Last evaluation | | | |
|---|---|---|---|---|---|---|
| | M | s | M | s | T | p |
| Substance abuse % | 100 | 0.0 | 36.00 | 41.0 | 6.74 | 0.0001 |
| Severity | 5.00 | 1.1 | 3.42 | 1.6 | 4.82 | 0.0001 |

Table 1 illustrates the T-Test results at baseline and endpoint. Alcohol intake at endpoint was approximately 36% with reduction of about 60%. Table 2 illustrates the percentage of alcohol intake in the sample breakdown. 9 patients had ceased drinking alcohol, 4 had not modified their drinking habit. The severity of illness at endpoint

**Table 2.** Alcohol intake (last assessment) in 20 depressed alcoholics treated with GHB and other psychoactive agents

| Alcohol intake % | N | % |
|---|---|---|
| 0.00 | 9 | 47.4 |
| 30.00 | 2 | 10.5 |
| 35.00 | 2 | 10.6 |
| 50.00 | 1 | 5.3 |
| 60.00 | 1 | 5.3 |
| 80.00 | 1 | 5.3 |
| 100.00 | 4 | 21.1 |

ranged between mildly and moderately ill. Both results are statistically significant. Subjects who are more impaired at baseline continue to be so at endpoint ($r=.54$ $p<.05$). Their global improvement is minor ($r=.82$ $p<.01$) and their efficacy index is worse ($r=.87$ $p<.01$). There is no relation between duration of treatment and improvement. This indicates that GHB is effective at the beginning of treatment.

GHB efficacy is reported in table 3. 38% of the sample showed an improvement. Only 1 subject's condition slightly worsened. In 3 subjects the clinical picture remained unchanged. Therefore 4 subjects are considered non responders. These non responders were not more severely ill at baseline (T value = -1.08 p=ns).

Side effects were observed in only 27.7% of the sample characterized by drowsiness lasting 1-2 hours. Vertigo was present in 50% of the subjects at the beginning of treatment. One subject presented a serious hyperammonaemia.

**Table 3.** Global improvement and side effects in 20 depressed alcoholics treated with GHB and other psychoactive agents

| | N | % |
|---|---|---|
| 1. Very much improved | 7 | 38.0 |
| 2. Much improved | 4 | 22.2 |
| 3. Minimally improved | 5 | 27.7 |
| 4. No change | 3 | 16.6 |
| 5. Minimally worse | 1 | 5.5 |
| 1. Side effects | 5 | 27.7 |
| 2. Somatic complications | 1 | 5.5 |
| 3. Psychopathological compl. | 0 | 0.0 |

Table 4 illustrates the efficacy index. 7 patients showed a net improvement with complete or near complete remission and no adverse effects. In 1 patient side effects interfered significantly with treatment. However the presence of side effects may be due to GHB, other psychoactive agents or the sum of the two.

Table 5 illustrates pharmacological agents and doses used. GHB is well tolerated in association with tricyclic and serotonergic antidepressants, mood stabilizers antihistamines and BDZ.

### B. Effectiveness and Tolerability in 4 Subjects Treated with Opioid Antagonists

We considered subjects previously treated with opioid antagonists and other psychoactive agents who had quickly (less than 1 week) relapsed in heroin use. They also had use of BDZ and/or alcohol secondary to craving for opioids. Exclusion criteria have been previously mentioned.

**Table 4.** Efficacy index in 20 depressed alcoholics treated with GHB and other psychoactive agents

| | | SIDE EFFECTS | | |
|---|---|---|---|---|
| **THERAPEUTIC EFFECT** | None | Do not significantly interfere | Significantly interfere | Outweighs therapeutic effect |
| **Marked.** Vast improvement. Complete or nearly complete remission of all symptoms | 38.0% | | | |
| **Moderate.** Decided improvement. Partial remission of symptoms | 5.5% | | | |
| **Minimal.** Slight improvement which doesn't alter status of care of patients | 22.2% | 22.2% | | |
| **Unchanged or worse.** | | 16.6% | 5.5% | |

**Table 5.** Associated pharmacotherapy and dosages used in 20 depressed alcoholics treated with GHB and other psychoactive agents

| | | Dosage | | | |
|---|---|---|---|---|---|
| | N paz | M | s | Min | Max |
| GHB | 20 | 24 | 9 | 10 | 60 |
| Imipramine | 4 | 92 | 33 | 50 | 150 |
| Clorimipramine | 4 | 36 | 16 | 20 | 50 |
| Trimipramine | 8 | 50 | 27 | 25 | 150 |
| Fluoxetine | 7 | 28 | 10 | 20 | 60 |
| Fluvoxamine | 11 | 130 | 66 | 50 | 300 |
| Paroxetine | 1 | 27 | | 20 | 40 |
| Carbamazepine | 6 | 510 | 92 | 400 | 800 |
| Acid valproic | 5 | 480 | 318 | 200 | 1000 |
| Sulpiride | 2 | 125 | 106 | 50 | 200 |
| Tiapride | 1 | 400 | | 400 | 400 |
| Clonazepam | 2 | 6 | 6 | 2 | 15 |
| Alprazolam | 1 | 100 | | 50 | 150 |
| Diazepam | 2 | 13 | 17 | 10 | 35 |
| Ademetil | 1 | 200 | | 200 | 200 |
| Phenobarbital | 1 | 100 | | 100 | 100 |
| Ciproeptadine | 1 | 12 | | 12 | 12 |

The data presented refer to 4 subjects, 2 males and 2 females, mean age 26 years (sd 6; min 17 max 32). 1 subject did not abuse alcohol. 3 subjects were moderately ill at baseline (CGI=4) and 1 was severely ill (CGI=6). Alcohol intake was greater than 2cc/Kg daily and the dose of BDZ used was more than 20mg/die. The duration of the observation period at endpoint was 43 days (sd 44; min 6 max 106). Six psychoactive agents were associated with GHB and Naltrexone. All patients were treated with a nightly dose of Trimipramine for insomnia.

Two of the three patients with a drinking problem ceased using alcohol and illicit BDZ; one subject reduced alcohol intake by 40%.The one patient who abused only BDZ ceased doing so. None of the 4 patients relapsed to heroin use. At endpoint 1 subject is minimally ill, 2 are mildly ill and 1 is moderately ill. There are no correlations between severity at baseline and at endpoint. Global improvement and efficacy index are proportional to duration of treatment (r.= .95 p<.05). The longer the patient stays in treatment the greater the clinical improvement. Efficacy and tolerability of GHB are evidenced by global improvement and efficacy index, in other words, the relation between therapeutic effect and side effects. One subject is much improved and 3 are minimally improved. No side effects were observed.

**Table 6.** Associated pharmacotherapy and dosages used in 4 subjects treated with opioid antagonists

| | | Dosage | | | |
|---|---|---|---|---|---|
| | N paz | M | s | Min | Max |
| GHB | 4 | 57.50 | 54.25 | 22.5 | 120 |
| Naltrexone | 4 | 50 | 0 | 50 | 50 |
| Carbamazepine | 1 | 400 | 0 | 400 | 400 |
| Lithium carbonate | 1 | 900 | 0 | 900 | 900 |
| Trimipramine | 4 | 100 | 0 | 100 | 100 |
| Fluoxetine | 1 | 40 | 0 | 40 | 40 |
| Imipramine | 1 | 75 | 0 | 75 | 75 |
| Fluvoxamine | 1 | 150 | 0 | 150 | 150 |

Table 6 illustrates psychoactive agents and doses used. GHB is well tolerated in association with other pharmacological agents and no side effects were evidenced.

### C. 2 Case Reports of Polydrug Abuse, Alcohol and BDZ in Undermedicated Heroin Addicts during Methadone Maintenance

V.D. is a 24 year-old male, unemployed polydrug abuser with a 10 year history of heroin addiction who has been in a Therapeutic Community. At the first consultation he presents psychomotor agitation due to abuse of Darkene (60cc/daily) and an unknown number of Zoplicone. In the last 6 months he has not used heroin nor opioid agonists.

agonists.

At baseline he was considered to be one of the most severely ill patients that we had seen with CGI=7. After 61 days of treatment he had reduced his use of BDZ by 75%. He was assessed as moderately ill (CGI=4) with consistent global improvement and no side effects.

Treated with GHB, mean dose 45 cc daily (min 30 max 80) GHB was associated with methadone, mean dose 30 mg.(min 30 max 35) and clonazepam, mean dose 4.75 mg (min 2 max 9).

F.M. is a 31 year-old female, unemployed, polydrug abuser with a 10 year history of heroin addiction. An outpatient of community public health services (SERT) treated with 10 mg daily of methadone, she is referred to us for alcoholic intoxication.

At ingress she is judged severely ill (CGI=6). After 24 days of treatment she had reduced alcohol intake by 70% and CGI score is 3, mildly ill with consistent global improvement and no side effects.

She is treated with GHB, mean dose 27 cc daily (min 20 max 30) GHB was associated with methadone, mean dose 27 mg.(min 10 max 30) and clonazepam, mean dose 4.75 mg (min 2 max 9). She was treated with a 100 mg/nightly dose of Trimipramine for insomnia.

## Conclusions

GHB appears to be effective as an anti-craving agent not only in alcoholic patients, as already demonstrated (1-3), but also in other drug addiction pathologies.

## References


1. Gallimberti L, Ferri M, Ferrara S D, Fadda F, Gessa G L. Gamma-hydroxybutyric acid in the treatment of alcohol dependence: A double-blind study. Alcohol Clin Exp Res 16:673-676, 1992.
2. Fadda F, Argiolas A, Melis M R, et al. . Differential effect of acute and chronic ethanol on dopamine metabolism in frontal cortex, caudate nucleus and substantia nigra. Life Sci 27:979-986, 1980.
3. Gessa G L. Nascita di un farmaco per l'alcoolismo: l'acido gamma-idrossibutirrico. In: Il GHB nel trattamento della dipendenza alcoolica, edited by Gallimberti L, Ferrarra SD, Gessa GL. Addiction Research Foundation of Italy, Padova, 1992, pp 3-4.

# 9

# Development of an Italian Version of the "Readiness to Change Questionnaire"[1] for Alcoholism and Addiction

**M. Scaglia, E. Baiardo, I. Baldasso, E. Cristina, S. Deodato, W. O. Faustman*, V. Spiller, M. Stefanini, R. Trampetti, G. P. Guelfi**
SERT-Servizio Tossicodipendenze USL N.3 "Genovese", Genova, Italy
* VA Medical Center and Stanford University School of Medicine, Palo Alto, CA, USA
(1) Translated with the permission of Dr. Steven Rollnick

## Summary

Prochaska and DiClemente developed a model for motivation to change addictive behavior. They classified patients into one of three stages as follows: (1) Precontemplation --the patient gives little or no consideration to reducing alcohol or drugs, (2) Contemplation --consideration to change is present but ambivalence is affecting change commitment, (3) Action -- the patient is willing to take active steps to reduce substance use. Rollnick et al. developed the "Readiness to Change Questionnaire" (RTCQ), a 12-item questionnaire that measures the degree to which a patient endorses the constructs of Precontemplation, Contemplation, and Action. We have translated the RTCQ into Italian and have performed initial psychometric studies (e.g. test/retest reliability) in our Italian clinic sample, a quite homogeneous sample of heroin-dependent patients. We found generally adequate (r=0.65 to 0.77) test-retest reliability estimates.

## Background

Motivation to change seems to be an important factor in the outcome of alcoholism and drug addiction, if one takes into account the natural course of such conditions or the response to the different forms of treatment.

The Prochaska and DiClemente model of change (1) describes the process of change underlying an attempt to resolve an addictive problem, both with and without treatment.

Since such model was first presented (2), its utility was acknowledged by workers in the field of alcoholism and drug addiction. Such article was recognized by a majority of them as the more influential for their job throughout the Eighties (3).

In its more recent formulation, the stage of change model of Prochaska and DiClemente (4) comprises a movement through the stages of Precontemplation,

Contemplation, Action, and Maintenance. From the stage of "precontemplation", in which dependence originates, characterized by the lack of acknowledgment of dependence itself as a problem, and in a way by the peaceful coexistence with it, the patient moves to the stage of "contemplation", in which the attitude toward dependence is dominated by the coexistence, markedly ambivalent, of the will of keeping on using the substance on one hand, and of the desire to quit it on the other. In previous versions, the stage of contemplation was followed by a phase of "determination", or decision making, that represented the accomplishment of its evolution toward the "Action" stage. In such stage, the patient tries to implement one or more of the methods potentially useful to actually overcome the process of dependence.

At this point of his evolution, the subject enters the stage of "maintenance", in which his detachment from the addictive behavior consolidates, through the adoption of a new life-style and definite treatment and rehabilitation programs, (i.e. relapse prevention). Relapse is considered an event, more than a stage, which can bring the patient back to the initial stages of the process.

The stages of change model has been shown to predict reliably outcome and drop-out from treatment for smoking cessation (4). Prochaska and DiClemente stages of change model is also potentially important because it can help workers in the field to chose the therapeutic approach: the allocation of the patient in the ladder of the model can be a factor of the choice of treatment.

Although thorough definitions are lacking, and more precise research are needed, Rollnick and colleagues (5) hypothesize that even the more efficient techniques of treatment for alcoholism and drug dependence are in fact irrelevant with patients in Precontemplation or in early Contemplation (that is, when they reject the acknowledgment of being out of control, or the pole of ambivalence in favor of quitting is still too weak). It is well known that such cases are very common, probably the majority of patients with dependence problems: such patients, in a frame of reference, are considered being "in denial". For patients in such early stages the techniques of "Motivational Interview" (6-7) can be accepted as the more appropriated, as they are primarily aimed at raising the level of the individual's motivation to undergo treatment. Motivational Interviewing is a unitarian approach "designed to elicit from the client increases in motivation to change addictive behavior, without the use of challenge confrontation" (5).

In this perspective, it is needed a psychometric instrument to assess patients' motivation to change, and a score that allows to allocate them to one of the Prochaska and DiClemente's stages. Such an instrument could be useful in the effort to predict the outcome of the intervention and in the decision on the treatment of choice.

## The Readiness to Change Questionnaire

Rollnick and coworkers (5) developed a questionnaire (Readiness to Change Questionnaire, RTCQ) aimed to allocate excessive drinkers to one of the Prochaska and DiClemente's stages of change. In their attempt they have kept in mind the need of an instrument brief, easy to administer, that could be used in non-specialistic settings, outside the field of alcohol and drug treatment facilities. In particular, they designed a Questionnaire that, differently from the `University of Rhode Island Change Assess-

**Table 1.** Readiness To Change Questionnaire (Miller et al. 1992)

| | |
|---|---|
| 1) | I don't think I drink too much. |
| 2) | I am trying to drink less then I used to. |
| 3) | I enjoy my drinking, but sometimes I drink too much. |
| 4) | Sometimes I think I should cut down on my drinking. |
| 5) | It's a waste of time thinking about my drinking. |
| 6) | I have just recently changed my drinking habits. |
| 7) | Anyone can talk about wanting to do something about drinking, but I am actually doing something about it |
| 8) | I am at the stage where I should think about drinking less alcohol |
| 9) | My drinking is a problem sometimes. |
| 10) | There is no need for me to think about changing my drinking |
| 11) | I am actually changing my drinking habits right now. |
| 12) | Drinking less alcohol would be pointless for me. |

Questionnaire that, differently from the `University of Rhode Island Change Assessment Scale (URICA)` (8) could be used by Precontemplators (ie, that would not imply any acknowledgment of the problem-dependency), and allowed the allocation with no further calculations to one of the stages of change (Table 1).

For some of the items, RTCQ is derived from URICA, and in the version we translated is build up of 12 items, each of them can be regarded as a scale measuring the extent to which the subject endorses that stage of change. Response points for items were deemed to run from -2 (strongly disagree) to +2 (strongly agree). The range of each scale was therefore -8 through 0 to +8. Three clusters of 4 items are expression respectively of the stages of Precontemplation, Contemplation, and Action. The questionnaire is designed to be self-administered, and its implementation requires few minutes.

The Italian version of the Questionnaire was modified, in comparison to the English version, in order to make it possible the administration to drug dependent patients, as opposed to alcohol dependent patients. Such adaptation implies some linguistic change, that could possibly affect the actual comparability of the two versions. To address this issue further study is required.

We have decided to work to an Italian version of RTCQ to import a simple instrument that could help us in decision making about the problems of the choice of treatment, ie, matching patients and treatment, a task we feel as critical in the field of alcoholism and drug addiction. The objective of the present study, therefore, was to verify the reliability of the RTCQ in the Italian contest, with a sample of subjects dependent on opiates attending our clinic in Genova.

## Methods

We have translated into Italian the Readiness to Change Questionnaire, and we have administered it to thirty-six patients attending our SERT in Genova.

*Subjects*

The majority of subjects were males (25 vs 11 females), the average age was 31.3 (+ 5.4), and the average duration of dependence on opiates was 10.5 years, (+4.4). A great majority (32) were in treatment with methadonc, some (2) were in treatment with opiate-antagonists, namely naltrexone.

**Table 2.** Reliability Test-Retest

| | |
|---|---|
| Precontemplation | r=0.70 (*) |
| Contemplation | r=0.81 (*) |
| Action | r=0.78 (*) |

(*) p< 0.001

The questionnaire was administered at the methadone center, while the patients were waiting for their turn; some subjects were recruited during the interview at the office.

*Data analysis*

We have performed a reliability-test of the RTCQ through a test-retest procedure.

**Table 3.** Correlation between the three scales

| | |
|---|---|
| Action/Contemplation $_{Test}$ | r= 0.10 (*) |
| Action/Precontemplation $_{Test}$ | r= -0.13 (*) |
| Precontemplation/Contemplation $_{Test}$ | r= -0.39 (**) |

(*) n.s.
(**) p< 0.5

## Results

Test-retest reliability: Pearson's correlation between the two occasions of administration of the questionnaire were as follows: Precontemplation = 0.70; Contemplation = 0.81; Action = 0.78, that represents satisfactory reliability for three four-items scales.

The correlation between duration (in years) of dependence and the values of the Precontemplation was r = .59 (p> .001). Such finding can be viewed as a first validity data, as it is consistent with the common view of increasing difficulties to change related to the duration of addiction.

The correlations between the scores of the three scale were calculated. An inverse correlation was found between Precontemplation and Contemplation scales.

The relative independence of the three scale suggests that they are actually measuring different, independent variables.

## Discussion

Test-retest reliability as results from our study is satisfactory, comparable to those obtained by Rollnick and colleagues (5) in their sample. Such similarity, despite the difference between a population of Welsh-Australian, English-speaking problem drinkers with low level of dependence, recruited in a general hospital setting, and Italian hard-core, opiate-dependent for long time, mainly treated with opiate agonist is likely to be highly significant of a real high reliability of the RTCQ.

The Italian version of RTCQ, as applied to a population of opiate-dependent subjects, in majority long-lasting methadone treatment, have demonstrated good psychometric properties, and a good reliability test-retest.

## References

1. Prochaska J O, DiClemente C C. Toward a comprehensive model of change. In: Treating Addictive Behaviors: Processes of Change, edited by Miller W R, Heather N. Plenum Press, New York, 1986,
2. Prochaska J O, DiClemente C C. Stages and process of self-change of smoking: toward an integrative model of change. J Consul Clin Psychol 51:390-395, 1983.
3. Davidson R. Prochaska and DiClemente's model of change: a case study? (Editorial). British Journal of Addiction 87:821-822, 1992.
4. Prochaska J O, DiClemente C C. Stages of change in the modification of problem behaviors. In: Progress in Behavior Modification, edited by Hersen M, Miller R M. SAGE, Newbury Park, Ca, 1992,
5. Rollnisk S, Heather N, Gold R, Hall W. Development of a short 'Readiness to Change' Questionnaire for use in brief, opportunistic intervention among excessive drinkers. British Journal of Addiction 87:743-754, 1992.
6. Miller W R. Motivational Interviewing with problem drinkers. Behavioral Psychotherapy 11:147-172, 1983.
7. Miller W R, Rollnick S. Motivational Interviewing: Preparing People to Change Addictive Behaviors, The Guilford Press, New York, 1991.
8. McConnaughy E A, Prochaska J O, Velicer W F. Stages of change in Psychotherapy: Measurement and sample profiles. Research and Practice 20:368-375, 1983.

# 10

# Mood Disorders in Heroin Dependence and Clinical Differences Between Heroin Addicts with and without Mood Disorders

**L. Daini[1,2], M. R. Capone[1], T. Agueci[1], M. Aglietti[1], O. Zolesi[1,2], I. Maremmani[1]**
[1]Dependence-Psychiatry Unit, Psychiatric Institute, University of Pisa, Italy
[2]Ph.D. Research Program Siena, Pisa & Cagliari Universities

**Summary**

Psychiatric comorbidity for mood disorders, investigated by means of a semistructured interview in 54 consecutive patients admitted to the Dependence-Psychiatry Unit of the Psychiatric Clinic of the University of Pisa, was found in 53 subjects when temperamental characteristics and family history were considered. A comparative study of the clinical characteristics of heroin addicts with and without mood disorders showed that there is a shorter latent period between first use and continued use of heroin in subjects with mood disorders. It would seem that mood disorders do not influence the course of drug addiction, which in our patients always reached the "revolving door" phase and that MMTPs are also needed in the case of psychiatric comorbidity. Our results show that in subjects suffering from mood disorders (primary or secondary) the transition to dependency occurs in a shorter period of time. These results indicate the importance of the early psychiatric diagnosis in heroin dependence.

The relationship between Mood Disorders and Heroin Dependence has been widely explored.

Until recently most researchers were primarily interested in identifying depressive symptomatology and defined substance abusers "depressed". These subjects were frequently assigned to diagnostic categories such as "recurrent depression" "single episode" "bipolar disorders". With the introduction of diagnostic criteria and systemic descriptions, as those of DSM for mental disorders, instruments have been developed that facilitate identification of diagnostic category subtypes by analyzing the relationship between principal diagnostic groupings and temperamental characteristics or family history. These instruments may help clarify the relationship between mood disorders and substance abuse and dependence (1-8).

These instruments are being used at the Dependence-Psychiatry Unit of Psychiatric Institute of the University of Pisa in the following studies: An epidemiological study for the identification of comorbidity for Mood Disorders in heroin addicts. A comparative study of clinical characteristics of heroin addicts with and without comorbidity for mood disorders.

In this paper preliminary data regarding these studies are presented. Characteristics of opioid dependence were investigated with the RSDA (Rating Scale for Drug Addiction) by Maremmani et al., (9). The RSDA is a heteroevaluation questionnaire divided into 6 areas which examine the following variables: physical condition (hepatic, vascular and lymph node pathology, gastroenteric disorders, sexual disorders, dental pathologies); mental state (awareness of illness, memory disorders, anxiety disorders, mood disorders, aggression, thought disorders, sensory perception disorders); socio-environmental conditions (employment family, sex, socialization and leisure time, legal problems); substance abuse (alcohol, opiates, CNS depressants, CNS stimulants, hallucinogens, phencyclidine, cannabis, inhalants, polysubstance abuse); substance abuse clinical picture (frequency of drug use, modality, pattern of use, phase, nosology; previous therapies; current therapy; methadone dosage; HIV positivity). The index card automatically provides 10 factors: physical problems, mental problems, polyabuse, previous treatments, combined treatments, occupational level, family situation, sexual problems, socialization and leisure time, drug related legal problems.

In order to evaluate mood disorders the SIMD (Structured Interview for Mood Disorders) by Cassano et al. was used (10). This is a semistructured interview for the diagnosis of depression appositely ideated for the identification of mood disorder subtypes on the basis of temperamental features, characteristics of course and interepisodic phases. After having diagnosed a Major Depressive Episode (DSM III R, 1987) and collecting data on course and family history a decisional tree permits the identification of 5 diagnostic subtypes: 3 Bipolar (I,II,III) and 2 Monopolar (Recurrent and Single Episode). Bipolar I subjects are those with a past history of manic episodes which meet DSM III R criteria; Bipolar II have past hypomanic episodes or premorbid hyperthymic temperament or cyclothymia, Bipolar III are defined on the basis of bipolar disorders in first degree relatives. The other two subgroups are identified on the basis of presence or absence of past depressive episodes respectively Recurrent Major Depression and Major Depression Single Episode.

## Epidemiology

40 heroin addicts are presently included in the study, 29 males (72%) and 11 female (28%) mean age, 25 years (sd. 3.9; min 19 max 37).

Table 1 illustrates preliminary results. Eight subjects (20%) met DSM III R diagnostic criteria for Major Depression Episode, have no past history for depression and are thus diagnosed as single episode. Three subjects (7.5%) have had a major depressive episode in the past and are in remission at the time of interview. One subject (2.5%) obtained a diagnosis of Recurrent Major Depression.

Major Depressive Episode is evidenced in another 15 cases (37.5%); in one case (2.5%) the presence of a manic episode permits the diagnosis of bipolar disorder; in the other 14 subjects (35%) the presence of hypomania spontaneous and/or induced by drugs or cyclothymic or hyperthymic temperament allows a diagnosis of Bipolar II to

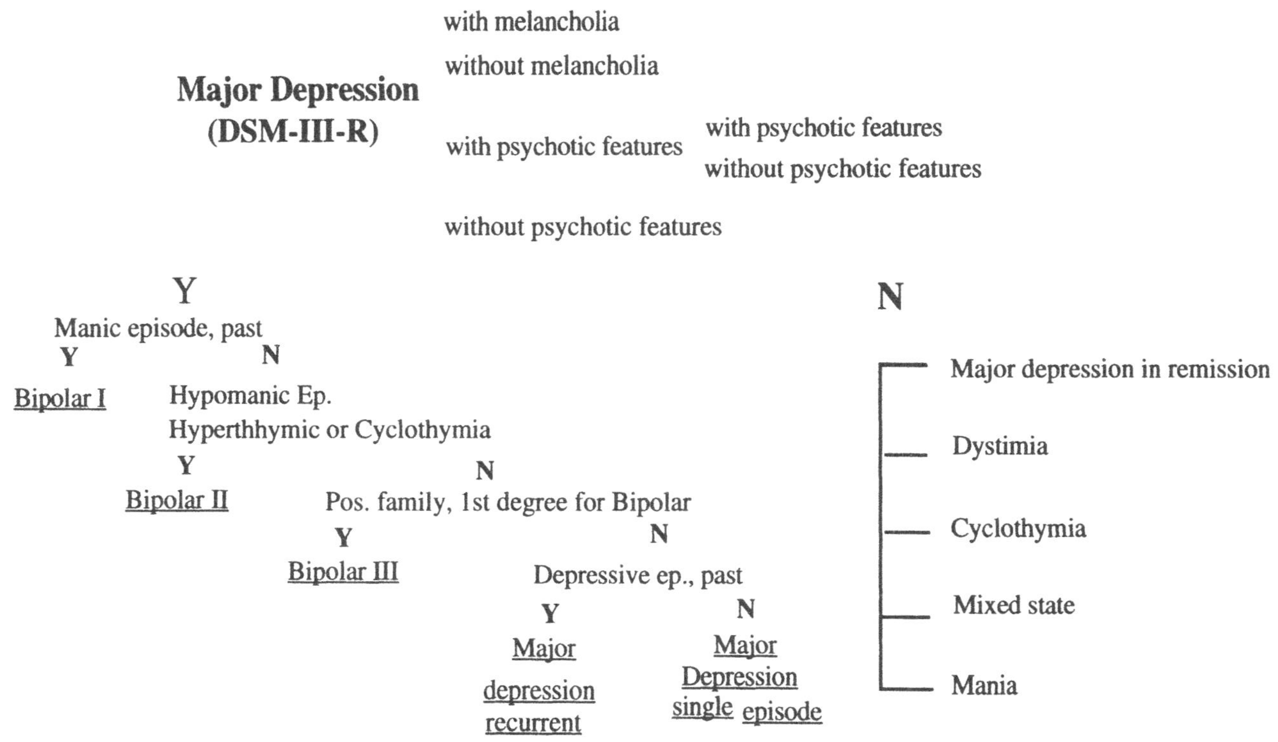


**Fig. 1.** SIMD (Structured Interview for Mood Disorders) by Cassano et al.

**Table 1.** Affective disorders in 40 heroin addicts

| | n | % |
|---|---|---|
| **Major depression** | | |
| Recurrent major depression | 1 | 2.5 |
| Major depression, single episode | 8 | 20.0 |
| Major depression, remission phase | 3 | 7.5 |
| **Bipolar disorders** | | |
| Bipolar I, depressive episode | 1 | 2.5 |
| Bipolar II, depressive episode | 14 | 35.0 |
| Bipolar III, depressive episode | 0 | 0.0 |
| Bipolar I, mixed episode | 0 | 0.0 |
| Bipolar I, manic episode | 0 | 0.0 |
| **Dysthymic disorder** | 1 | 2.5 |
| **Cyclothymic disorder** | 1 | 2.5 |
| **Psychotic episode** | | |
| Depressive episode with congruous manifestations | 0 | 0.0 |
| Depressive episode with incongruous manifestations | 2 | 5.0 |

be made. On the whole, affective disorders are present in 67.5% of the sample with a strong predominance of bipolar disorders.

Although a major depressive episode is relatively frequent (67,5% of subjects), a depressive disorder may be diagnosed in only 2.5% of the sample since the evolution of single episodes remains unknown. More representative appear to be bipolar disorders.

The values found in literature with reference to depressive disorders are about 73% and refer only to Major Depressive Episode. The diagnostic subtypes obtained with the SIMD lean more toward bipolarity.

Extremely rare are Dysthymia and Cyclothymia each found in 1 subject (2.5%). Psychotic episodes are an exception regardless of the polarity of the affective disorder. Only 2 subjects manifested psychotic features incongruous with mood. One subject presented melancholia.

If we also consider Cyclothymia and Dysthymia the affective disorder load in heroin dependence reaches 72.5%.

A closer look at temperament revealed that 35 of 40 (87.5%) subjects presented modifications: hyperthymic (n=25 62.5%) or depressive (n=10 25%). However, only in 15 of 25 subjects with hyperthymic temperament is a mood disorder present, whereas all 10 subjects with depressive temperament have an affective disorder. Therefore, 10 subjects (25%) with a hyperthymic temperament do not present mood modifications identifiable with a major disorder.

These subjects, on the one hand, increase the incidence of mood disorders in heroin dependence to 97.5%. On the other hand, they reinforce the importance of the bipolar spectrum.

The SIMD verifies the importance of psychiatric family history (Table 2). Psychiatric disorders in first or second degree relatives are present in 28 subjects (70%). In addition to positive family history 12 of these subjects present hyperthymic or depressive temperament without an affective disorder. The other 16 subjects with positive family history present a mood disorder. Of the 12 remaining subjects with a negative family history, 2 have temperamental features without an identifiable mood disorder, 8 have a mood disorder, 1 has a mood disorder with temperament changes. Only 1 of 40 subjects has neither affective disorders, temperament modifications nor positive family history (Table 3).

### Clinical Differences

54 heroin addicts are presently included in the study, 39 males (73,6%) and 14 female (26,4%) mean age 26 years (sd. 4; min 19 max 37).

Comorbidity for Mood Disorders was found in 35 subjects (83,3% males; mean age 26 years). The remaining 18 subjects (83,0% males; mean age 26 years) had neither affective disorders, temperament modification nor positive family history.

Tables 4-5 show the results. Although no differences were found between the two groups regarding clinical and toxicological characteristics, subjects with MD have more organic complications even though this is not statistically significant.

Social functioning on the whole does discriminate the two groups. As in literature, we found that where psychiatric comorbidity is present social functioning is more impaired. Within the area of social functioning, occupational activity and leisure time are problematic for the depressed addict.

There are no differences between the two groups as far as polydrug abuse goes. Nor in previous treatments undergone. Statistical significance is present for current treatments. Heroin addicts with comorbidity require more therapeutic attention and more

**Table 2.** Familial Psychiatric History in 40 heroin addicts

| | n | % |
|---|---|---|
| **1st Degree** | | |
| Major depression | 9 | 22.5 |
| Bipolar disorder | 6 | 15.0 |
| Alcoholism | 3 | 7.5 |
| Drug Abuse | 6 | 15.0 |
| Generalised anxiety disorder | 2 | 5.0 |
| **2nd Degree** | | |
| Major depression | 8 | 20.0 |
| Bipolar disorder | 1 | 2.5 |
| Alcoholism | 7 | 17.5 |
| Drug Abuse | 4 | 10.0 |
| Schizophrenia | 1 | 2.5 |
| Generalised anxiety disorder | 1 | 2.5 |
| Panic attacks disorder | 1 | 2.5 |

**Table 3.** Summary table

| | Temperament | | | |
|---|---|---|---|---|
| | Absence | | Presence | |
| | n | % | n | % |
| Affective disorders with positive family history | 16 | 40.0 | | |
| Affective disorders without positive family history | 9 | 22.5 | | |
| No affective disorders with positive family history | | | 12 | 30.0 |
| No affective disorders without positive family history | 1 | 2.5 | 2 | 5.0 |

**Table 4.** Comparison of clinical characteristics of heroin addicts

| | Gr=1 Without MD n=19 | | Gr=2 With MD n=35 | | | |
|---|---|---|---|---|---|---|
| | M | s | M | s | T | p |
| Age 1st use | 18 | 2 | 18 | 3 | .96 | .34 |
| Age continued use | 22 | 4 | 19 | 3 | 2.11 | .044 |
| Duration (months) | 63 | 64 | 76 | 58 | -.65 | .52 |
| Age 1st treatment | 23 | 3 | 22 | 4 | 1.74 | .08 |
| Organic disease | 1.31 | 1.1 | 1.77 | 1.3 | -1.28 | .20 |
| Psychopatology | 2.42 | 1.5 | 2.48 | 1.8 | -0.14 | .89 |
| Social adjustment | 2.15 | 1.6 | 2.97 | 1.6 | -1.75 | .08 |
| Work | .31 | .4 | .74 | .5 | -3.07 | 0.004 |
| Family | 1.21 | 1.6 | 1.02 | .8 | .46 | .65 |
| Sex | .26 | .4 | .34 | .4 | -.60 | .54 |
| Leisure | .42 | .5 | .82 | .7 | -2.44 | 0.018 |
| Legal problems | .47 | .7 | .71 | 1.0 | -0.97 | .33 |
| Polyabuse | 4.05 | 1.3 | 4.08 | 1.9 | -.07 | .94 |
| N. Previous treatmets | 1.36 | .9 | 1.57 | 1.3 | -.64 | .52 |
| N. current treatments | 1.05 | .6 | 1.62 | .4 | -3.49 | 0.001 |

**Table 5.** Comparison of clinical characteristics of heroin addicts

| | Gr=1 Without MD n=19 | | Gr=2 With MD n=35 | | | |
|---|---|---|---|---|---|---|
| | n | % | n | % | chi | p |
| Substance abuse | | | | | | |
| Alcohol | 9 | 50.0 | 19 | 55.9 | .16 | .68 |
| SNC Depressants | 7 | 36.8 | 16 | 47.1 | .51 | .47 |
| SNC Stimulants | 12 | 63.2 | 20 | 58.8 | .09 | .75 |
| Hallucinogens | 6 | 31.6 | 14 | 41.2 | .47 | .48 |
| Cannabis | 19 | 100.0 | 31 | 88.6 | 2.34 | .12 |
| Inhalants | 0 | 0.0 | 2 | 5.7 | 1.12 | .28 |
| Polyabuse | 7 | 36.8 | 15 | 44.1 | .26 | .60 |
| Poliabuse & Methadone | 5 | 26.3 | 6 | 17.6 | .55 | .45 |
| Clinic Characteristics | | | | | | |
| Heroin intake | | | | | | |
| Sporadic | 2 | 10.5 | 4 | 11.4 | | |
| Weekly | 0 | 0.0 | 3 | 8.6 | | |
| Daily | 17 | 89.5 | 28 | 80.0 | 4.15 | .24 |
| Typology | | | | | | |
| Stables | 11 | 57.9 | 11 | 31.4 | | |
| Junkies | 5 | 26.3 | 14 | 4.0 | | |
| Two wonders | 2 | 10.5 | 5 | 14.3 | | |
| Loners | 1 | 5.3 | 5 | 14.3 | 3.80 | .28 |
| Periodic abstinence | 14 | 82.4 | 30 | 88.2 | .33 | .56 |
| Phase of illness | | | | | | |
| Honey moon | 1 | 5.6 | 5 | 14.7 | | |
| Dose increase | 4 | 22.2 | 6 | 17.6 | | |
| Revolving door | 13 | 72.2 | 23 | 67.6 | 1.01 | .60 |
| Nosography | | | | | | |
| Reactive | 1 | 6.7 | 6 | 17.1 | | |
| Selfmedicating | 1 | 6.7 | 5 | 14.3 | | |
| Metabolic | 13 | 86.7 | 24 | 68.6 | 1.79 | .40 |

pharmacotherapy).

Although the age of the first contact with heroin is not different for these two groups, the age of continued drug use is minor for the group with comorbidity.

No differences are evident between the two groups for history of drug abuse. Most subjects had a daily pattern of use, had periodic drug free states and relapses. Both groups were "metabolic drug addicts" according to Maremmani and Castrogiovanni classification (11).

**Discussion**

The affective disorder load in our patients is quite consistent (72.5%) and reaches 97.5% if we consider hyperthymic temperament. In fact, all subjects with temperament without an affective disorder are hyperthymic. Contrary to many other studies (8) (1) (4) (12), bipolar and expansive pathologies seem to be strongly represented (2.5% Depressive Episode in Bipolar I; 37.5% Bipolar II; 2.5% Cyclothymia) when compared to Unipolar (2.5% Dysthymia, 2.5% Major Recurrent Depression). Additional data could be derived from a longitudinal study of single depressive episodes (27.5%) which presently do not permit us to exclude bipolarity. If we consider Bipolar II and hyperthymia the expansive polarity appears closely linked to opioid dependence (67.5%). This observation is in accordance with the frequently reported substance use in Bipolar Disorder, Cyclothymia and in the expansive phases of illness. Moreover, bipolar subjects abuse more than one substance; consistently the patients in our study leaned toward the bipolar spectrum and were polysubstance abusers (13). Therefore, mood disorders evidenced in our sample belong to the bipolar spectrum even though most patients were depressed at the time of the interview. In agreement with the literature, our depressed addicts have a high level of education and respond favorably to treatment (7,14,15). About 70% are undergoing differential treatment (long term with agonists, long term with antagonists with or without antidepressants). Finally, family history (70%) strongly influences heroin addiction. All subjects with a positive family history present an affective disorder or hyperthymic or depressive temperament; noteworthy is the finding of 1 subject who had neither a mood disorder, temperament modification nor positive family history. Heuristically these data support the hypothesis of a primary mood alteration which leads to substance abuse which per se then causes dependency. Consequently, drug dependence through self-medication would appear to be re-evaluated (16) (17).

Affective pathology frequently present in substance abusers does not seem to influence clinical characteristics of illness. Comorbidity for mood disorders, in drug addicts, influences social functioning, and although both groups begin drug use at the same age the group with comorbidity reaches dependence in a shorter period of time. Drug use whatever the cause in time becomes an independent pathology, thus requiring specific treatment. Accurate diagnosis is important in those addicts with affective disorders especially since these subjects reach dependence in a shorter period of time.

**References**


1. Dorus W, Senay E C. Depression demographic dimensions, and drug abuse. Am J Psychiatry 137:669-704, 1980.
2. Lehman W X, De Angelis G C. Adolescents, methadone, and psychoterapeutic agents. In: Proceedings of the Fourth National Conference on Methadone Treatment, National Association for the Prevention of the Addiction to Narcotics, New York, 1972,
3. Robbins P R. Depression and drug addiction. Psychoanal Quart 48:375-386, 1974.
4. Rounsaville B J, Weissman M M, Crits-Christoph K, Wilber C H, Kleber H B. Diagnosis and symptoms of depression in opiate addicts: Course and relationship to treatment outcome. Arch

5. Steer R A, Kotzer E. Affective changes in male and female methadone patients. Drug and Alcohol Dependence 5:115-122, 1980.
6. Weiland W F, Sola S. Depression in opiate addicts measured by objective tests. In: Proceedings of the Third National Conference on Methadone Treatment, National Association for the Prevention of Addiction to Narcotics, New York, 1970,
7. Weissman M M, Slobetz F, Prusoff B A, Mesritz M, Howard P A T. Clinical depression among narcotic addicts maintained on methadone in the community. Am J Psychiatry 133:1434-1438, 1976.
8. Rounsaville B J, Kosten T R, Weissman M M, Kleber H D. Evaluating and Treating Depressive Disorders in Opiate Addicts, NIDA National Institute on Drug Abuse, Rockville, Maryland, 1985.
9. Maremmani I, Castrogiovanni P. Ab.So.: Scheda per la rilevazione dell'abuso di sostanze, Università di Pisa, Istituto di Clinica Psichistrica, 1989.
10. Cassano G B, Akiskal H A, Musetti L, Perugi G, Soriani A, Mignani V. Psychopathology, temperament and paste course in 405 Primary Major Depression explorations with a New Semi-Structured Interview for Depression. Psychopathology 1992.
11. Maremmani I, Castrogiovanni P. Disturbi da uso di sostanze. Disturbi da Oppiacei ed Analgesici. In: Trattato Italiano di Psichiatria. Volume 2 (43), edited by Cassano G B, D'Errico A, Pancheri P, Pavan L, Pazzagli A, Ravizza L, Rossi R, Smeraldi E, Volterra V. Masson, Milano, 1992, pp 1148-1161.
12. Robbins E, Prosch W A, Stern M. Further observations on untoward reactions to LSD. Am J Psychiatry 124:393-395, 1967.
13. Miller F T, Busch F, Tanenbaum J H. Drug abuse in schizophrenia and bipolar disorders. Amer J Drug Alcohol Abuse 15(3):291-295, 1989.
14. Titievsky J, Seco G, Barranco M, Kyle E M. Doxepin as adjunctive therapy for depressed methadone maintenance patients: A double blind study. J Clin Psychiat 43:454-456, 1982.
15. Kleber H D, Weissman M M, Rounsaville B J, Prusoff B A, Wilbur C H. Imipramine as treatment for depression in opiate addicts. Arch Gen Psychiatry 40:649-653, 1983.
16. Wurmser L. Psychoanalytic considerations of the etiology of compulsive drug abuse. American Psychoanalitic Association 22:820-843, 1974.
17. Khantzian E J. An ego/self theory of substance dependence. In: Theories on Drug Abuse, edited by Lettieri D J, Sayers M, Pearson H W. NIDA Research Monograph 30, Washington,D.C.: Supt. of Docs, U.S. Govt. Print. Off., 1980,

# 11
# Nature and Magnitude of Problems Associated with Psychoactive Drugs

**A.H. Ghodse, S. Rawaf**
Division of Addictive Behaviour, St George's Hospital Medical School, London, U.K.

## Summary

It is an unfortunate fact that many prescribed psychoactive drugs are drugs of abuse - that is to say they are drugs that are used excessively in a way that is inconsistent with or unrelated to acceptable medical practice. Of course, they are not alone in this and the use of prescribed drugs must be placed in perspective as one part of a spectrum of drug abuse behaviour. On one side are illicit drugs such as heroin, cocaine, LSD and cannabis; controlling their availability clearly comes within the province of law enforcement agencies. On the other side are drugs, both medicinal and recreational, which are available without prescription; responsibility for controlling the availability of the two most important recreational drugs, tobacco and alcohol, resides with government. In the middle are the prescribed psychoactive drugs - drugs such as hypnotics, tranquilizers, sedatives, stimulants etc.-, and although available only on prescription, they are the subject of widespread abuse and cause serious public health and social problems. Since direct measures of the prevalence of drug abuse problems are not available, other indicators are used instead. Prominent among these are studies of morbidity - the problems that arise as a consequence of drug abuse. Apart from acting as indicators of drug abuse, such problems are worthy of study in their own right, permitting the development of more effective health and social services, and of better preventative measures.

## Nature and Magnitude of the Problem

Because of intensive testing by the pharmaceutical companies, most of the side effects of prescribed psychoactive drugs are well known and it is not the purpose of this paper to enumerate the blood dyscrasias, skin rashes, cardiac arrhythmias etc., that may occur with different drugs. Instead it will concentrate on those problems that occur in the context of widespread use and misuse and which lie at the centre of current concern.

It is an unfortunate fact that many prescribed psychoactive drugs are drugs of abuse - that is to say they are drugs that are used excessively in a way that is inconsistent with, or unrelated to, acceptable medical practice. Of course, they are not alone in this and the use of prescribed drugs must be placed in perspective as one part of a spectrum of drug

**Table 1a.** Number of prescription by therapeutic groups and selected sub-groups

| Great Britain | Millions | | | | | | | | | |
|---|---|---|---|---|---|---|---|---|---|---|
| | 1980 | 1982 | 1983 | 1984 | 1985 | 1986 | 1987 | 1988 | 1989 | 1990 |
| All groups | 361 | 370 | 376 | 382 | 380 | 384 | 400 | 413 | 421 | 431 |
| **Cental Nervous System** | **84** | **84** | **81** | **82** | **76** | **77** | **80** | **81** | **81** | **85** |
| Hypnotics | 17 | 17 | 16 | 16 | 16 | 16 | 17 | 16 | 15 | 15 |
| Sedatives and tranquillisers | 22 | 21 | 18 | 16 | 14 | 14 | 13 | 12 | 11 | 10 |
| Anticonvulsants | 4 | 4 | 4 | 4 | 5 | 5 | 5 | 5 | 5 | 5 |
| Anti-parkinsonism drugs | 2 | 2 | 2 | 2 | 2 | 2 | 2 | 2 | 2 | 2 |
| Analgesics major | 4 | 5 | 5 | 5 | 5 | 5 | 6 | 6 | 6 | 6 |
| Analgesics minor | 21 | 21 | 20 | 22 | 20 | 20 | 23 | 25 | 26 | 28 |
| Antidepressants | 7 | 7 | 7 | 7 | 7 | 8 | 8 | 8 | 9 | 9 |

abuse behavior. On one side are illicit drugs such as heroin, cocaine, LSD and cannabis; controlling their availability clearly comes within the province of law enforcement agencies. On the other side are drugs, both medicinal and recreational, which are available without prescription; responsibility for controlling the availability of the two most important recreational drugs, tobacco and alcohol, resides with government. In the middle are the prescribed psychoactive drugs - such as hypnotics, tranquillizers, sedatives, stimulants etc. Although these are available only on prescription, they are the subject of widespread abuse and cause serious health and social problems.

This is not really surprising when one considers the sheer number of prescriptions that are issued for these drugs. For example, in the United Kingdom alone, more than 85 million prescriptions for psychoactive drugs were issued by primary health care practitioners in 1990 (Table 1) (1).

The scale of prescribing cannot, however, be adequately represented merely by the number of prescriptions issued. There may, for example, be many prescriptions for small quantities of drugs, or indeed many prescriptions for large quantities. For this reason, data on the total production as well as on the consumption of drugs are important guides. Figures 1 and 2 show the production in one year of 2 major classes of benzodiazepines. Figure 3 shows trends in the production and utilization of methadone to give another example. The method adopted by WHO, expressing drug consumption in terms of defined daily dose (DDD) per year per one thousand population gives a better picture and is a much better indicator of drug utilization. Table 2 shows the therapeutic use of various psychoactive drugs in Nordic countries in 1980. It seems unlikely that the Danes really suffer so much more from insomnia than the Finns, or that they are really so much more anxious. These figures, therefore, illustrate very clearly how prescribing practices can vary, and emphasize the need for rational use of drugs by all health care professionals (2).

**Table 1b.** Number of prescription by therapeutic groups and selected sub-groups

| | % change: | |
|---|---|---|
| | **1980-90** | **1970-80** |
| **All groups** | **19** | **22** |
| **Cental Nervous System** | **2** | **4** |
| Hypnotics | 12 | 27 |
| Sedatives and tranquillizers | 54 | 19 |
| Anticonvulsants | 22 | 14 |
| Anti-parkinsonism drugs | 33 | ---- |
| Analgesics major | 41 | 328 |
| Analgesics minor | 33 | 7 |
| Antidepressants | 30 | 5 |

Tables 3, 4 and 5 demonstrate the consumption of 3 drugs and are examples of how practices vary in different parts of the world (3).

One of the most important morbidities associated with the abuse of psychoactive

**Table 2.** Therapeutic use of psychoactive substances in various European Countries in 1980, expressed as defined daily doses (DDD) per 1000 adults
(source: Ghodse A.H. & Khan I., 1988)

| Category | DK | SF | IC | N | S |
|---|---|---|---|---|---|
| Neuroleptics | 9.63 | 9.68 | 5.91 | 8.42 | 10.26 |
| Tranquillizers | 40.78 | 19.63 | 24.23 | 22.64 | 21.86 |
| Hypnotics | 75.81 | 17.15 | 30.15 | 41.89 | 41.36 |
| Antidepressants | 10.18 | 4.05 | 11.25 | 8.51 | 7.75 |
| Stimulants | 1.01 | 0.89 | 0.67 | 0.18 | 0.69 |

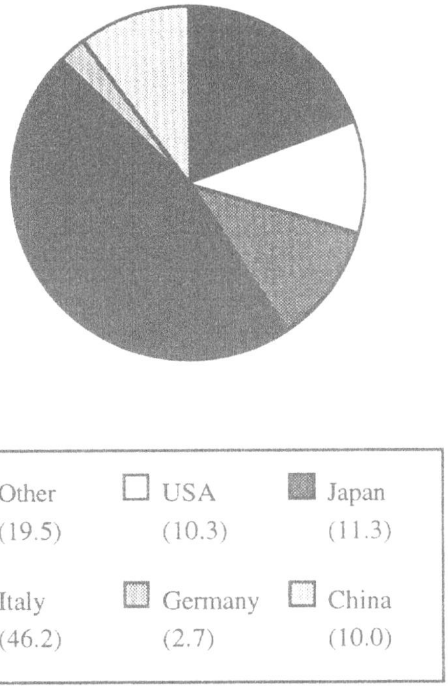


**Fig.1.** Benzodiazepine-type sedative hypnotics
Countries' share in total reported manufacture in 1991
(3 746 million DDD)

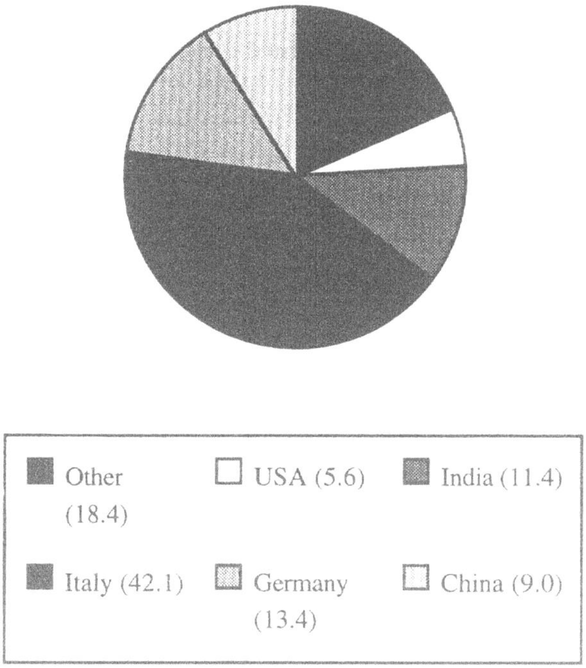


**Fig. 2.** Benzodiazepine-type tranquillizers
Countries' share in total reported manufacture in 1991
(3 746 million DDD)

drugs is that of overdose. In numerical terms it is a vast problem. In London, for example, in the mid-1970s the annual prevalence rate, almost certainly an underestimation, was calculated to be about 430 per 100,000 of the population, meaning that one person in 230 of the adult population aged between 15 and 65 years took an overdose each year (4) - and the majority of these overdoses involved one or more psychoactive drugs. Similar prevalence rates occur in other countries and it must be emphasized that it is not just an interesting statistic: many people put their lives at risk by taking a drug overdose. Indeed such is the frequency of this form of drug abuse that the margin of safety between the therapeutic dose required for serious overdose or death is often a property of commercial significance. This is undoubtedly one factor contributing to the success of benzodiazepines in the market place as they are so much safer when taken in overdose than the barbiturates which they largely replaced. As a result mortality due to drug overdose declined in many places even though this type of drug abuse behavior has not.

As well as being drugs of abuse many psychoactive drugs also cause dependence - the compulsion to take them on a continuous or periodic basis in order to experience their psychic effect and sometimes to avoid the discomfort of their withdrawal (or abstinence) syndrome. It is this phenomenon of dependence that lies at the core of anxiety about the

**Table 3.** The Worldwide Consumption of Methaqualone 1989/91

| | Number of defined daily doses (DDD) per thousand inhabitants per day |
|---|---|
| Austria | 1.39 |
| Belgium | 1.34 |
| Switzerland | 1.25 |
| Spain | 0.07 |
| France | 0.05 |
| Portugal | 0.04 |
| Netherlands | 0.02 |
| Denmark | 0.01 |

**Table 4.** Number of defined daily doses (DDD) per thousand inhabitants per day

| | Number of defined daily doses (DDD) per thousand inhabitants per day |
|---|---|
| France | 5.49* |
| Falkland Islands | 2.95 |
| Switzerland | 2.40 |
| Israel | 2.29 |
| U.S.A. | 1.38 |
| Cuba | 0.72 |
| Belgium | 0.46 |
| British Virgin Islands | 0.36 |
| Chile | 0.28 |
| Iceland | 0.27 |
| Spain | 0.26 |
| U.K. | 0.18 |
| Sweden | 0.07 |
| Canada | 0.06 |
| Denmark | 0.06 |
| Norway | 0.05 |
| Australia | 0.04 |
| Luxembourg | 0.04 |
| Netherlands | 0.03 |
| French Polynesia | 0.02 |
| Malta | 0.02 |

* Includes mainly conversion into other substances

**Table 5.** Medical use of methylphenidate 1989/91

| | Number of defined daily doses (DDD) per thousand inhabitants per day |
|---|---|
| U.S.A. | 0.86 |
| Canada | 0.43 |
| Cayman Islands | 0.24 |
| Israel | 0.17 |
| Norway | 0.17 |
| Switzerland | 0.10 |
| Australia | 0.07 |
| Cuba | 0.07 |
| Chile | 0.07 |
| Iceland | 0.07 |
| British Virgin Islands | 0.06 |
| Mexico | 0.05 |
| Japan | 0.05 |
| South Africa | 0.05 |
| Ecuador | 0.05 |
| Bahamas | 0.04 |
| Panama | 0.04 |
| Hong Kong | 0.04 |
| Belgium | 0.04 |
| Denmark | 0.03 |
| New Zealand | 0.03 |
| Costarica | 0.03 |

**Table 6.** Depressants. Total quantities seized worldwide 1976-1986

| Yearly Average | Depressant | |
|---|---|---|
| **Year** | **Kg** | **Units** |
| 1976 | 12 | 1 366 514 |
| 1977 | 2 127 | 1 036 434 |
| 1978 | 3 861 | 517 672 |
| 1979 | 117 | 12 412 778 |
| 1980 | 2 654 | 25 324 609 |
| 1981 | 24 090 | 23 036 295 |
| 1982 | 1 656 | 12 773 238 |
| 1983 | 512 | 3 771 033 |
| 1984 | 2 069 | 4 679 949 |
| 1985 | 1 041 | 7 031 535 |
| 1986 | 2 486 | 6 810 545 |

abuse of psychoactive substances. The way in which it develops is not understood, but it seems to be closely related to what is described as the reinforcing properties of the drug. This is demonstrated most easily in a laboratory situation where an animal obtains a dose of a drug, such as cocaine, by pressing a lever. Thereafter, the animal presses the lever repeatedly to obtain more cocaine and as it receives more, presses the lever more and more rapidly to get still more. In other words, cocaine increases - or reinforces - behaviour resulting in its own administration and is said to be a primary reinforcer and to have primary reinforcing properties. It is not alone in possessing this property. Other drugs such as amphetamines, opiates, sedative-hypnotic and alcohol do too, although not all are such powerful primary reinforcers as cocaine. It is not known why drugs with otherwise very different pharmacological properties should share the property of primary reinforcement, but they do, and the consequent problems are intensified by the fact that many of these drugs also induce tolerance. This is a state of reduced responsiveness to the effects of a drug, caused by its previous administration. It is as if the body gets used to the drug so that in order to maintain its effect, the dose has to be increased and when the dose is increased, dependence becomes progressively more severe. This is manifested by the signs and symptoms of the abstinence syndrome if the drug is withdrawn, and it is worth pointing out that the abstinence syndrome for some drugs may be of life-threatening severity. The abstinence syndrome due to sedative-hypnotic drugs, for example, may cause a psychotic condition or status epilepticus. Other drugs may be less dangerous on withdrawal although the patient's symptoms may still be very distressing.

It is obvious that in a society where psychoactive drugs are widely available, many individuals who are vulnerable to dependence are likely to be exposed to the drugs and many will take them in an uncontrolled fashion. This involves not just the so called "hard" drugs (e.g. heroin, other opiates, cocaine), but also prescribed psychoactive drugs. Most recently dependence on benzodiazepines has been causing great concern, but in past decades barbiturates and other drugs have been the focus of attention. Many people have become dependent unwittingly in the course of medical treatment; while they continue to take their drugs they may be unaware that they are dependent on them, but as soon as they stop taking them unpleasant symptoms develop. In the case of benzodiazepine dependence, symptoms such as anxiety, tremulousness etc., are often so similar to the original symptoms for which the drugs were prescribed that patients believed that they are not yet "cured" and continue taking their medication - thus perpetuating the dependent state.

Unfortunately, the problem of dependence is not restricted to those for whom the medication was originally prescribed. Where these drugs are widely and excessively prescribed they contribute to a vast pool of drugs within the community as a whole. They become incorporated into daily life in a similar way to tobacco and alcohol, if not to quite the same extent yet. Already they are shared and borrowed like cigarettes and are taken like alcohol and often with alcohol, to achieve a relaxed mental state, providing a never-ending source of drugs for abuse and dependence just as surely as do the fields of opium poppies and the plantations of coca bushes and cannabis. This is recognized by the Commission on Narcotic Drugs which monitors illicit traffic of psychotropic drugs as well as seizures of opium, cannabis and cocaine. Table 6 illustrates the level of diversion of CNS depressant drugs.(U.N. Statistics 1989). Obviously these figures, which rely on

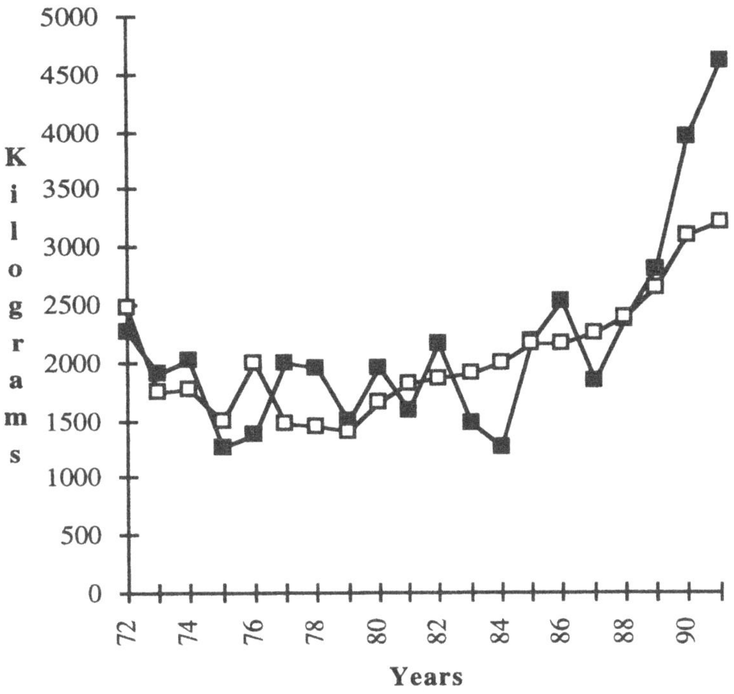


**Fig. 3.** Methadone: global manufacture (black pointer) and consumption (white pointer)

information communicated by governments in their annual reports, cannot be taken as entirely reliable statistics, but they are clear evidence of the diversion of drugs that should only be available on prescription.

The effects of drug abuse and dependence may be tragic for the individual concerned, in that it adversely affects their health and indeed may cause their death, but it also has profound implications for the wider society. Illicit drug production corrupts the economies of some countries and drug-related crime seriously disrupts life in some cities. Less newsworthy are the more subtle effects of the use and abuse of prescribed psychoactive medication -impairment of performance at work, increased incidence of accidents at work and on the road and so on. It is impossible even to guess at the contribution of psychoactive medication to incidents such as these, but given the vast scale of prescription of these drugs one fear that it is substantial.

In addition, to the social and economic consequence of drug abuse and dependence, serious public health problems may also ensue. These are closely related to the practices of drug abusers who inject their drugs - whether intramuscularly or intravenously. Whichever route is chosen injection carries a high risk of infection because sterilization, if attempted at all, is usually inadequate and because syringes and needles are often shared. Local infections -hepatitis, abscesses etc - are therefore common, but from a public health point of view the serious infections are those which are transmitted directly from one drug abuser to another by infected blood contaminating shared injection equipment. The most serious diseases transmitted this way are syphilis, hepatitis and AIDS. All are transmitted between injecting drug abusers in the manner described, but can also be transmitted sexually to non-drug abusing individuals. Thus drug abusers form a crucial route for these infections into the general population. This has long been the case for syphilis and hepatitis, but because they are familiar and not usually fatal, concern on the part of the general population has been muted. With AIDS which has a very high mortality rate, the situation is quite different and there is now a general consensus that drug abuse in all its forms, must be tackled as a significant preventative measure against AIDS.

The risk of AIDS may seem remote from the issues of the rational use of psychoactive drugs, most of which are prescribed and taken orally. However, the following points should be borne in mind.

Firstly, the altered-mental state consequent on drug abuse make casual sexual relationships (and the risk of AIDS) much more likely. Next, many drugs intended and prescribed for oral use may be crushed and then injected with all the attendant risks. Finally, and probably most important of all, the easy availability of prescribed psychoactive substances engenders an attitude that makes all forms of drug taking much easier and more acceptable. Thus where there is total background acceptance of drug taking behaviour, it is only a small step to much more harmful forms of drug abuse with a greater risk of AIDS.

Another serious problem associated with the abuse of psychoactive drugs is the effect on the fetus and new born child in cases of maternal drug abuse. This has been a cause of concern for nearly a century, when maternal addiction to morphine was the main problem, but a much wider range of drugs is now involved. Although taking drugs during pregnancy is a violation of the general rule that no unnecessary substances should be taken at this time, none of the common drugs of abuse has been implicated as a specific teratogen. Alcohol, however, which is usually one component of the polydrug abuser's cocktail, does have teratogenic effects. Further problems arise if the mother takes dependence-producing drugs regularly during pregnancy. Many of these drugs cross the placenta so that the fetus is constantly exposed to them too and also becomes dependent on them, suffering withdrawal if the mother is deprived of her drug. This may precipitate fetal distress or death or induce premature labour. Alternatively, if the mother takes a drug overdose, this may also affect the fetus adversely. Maternal dependence on sedative-hypnotic drugs is particularly hazardous because it is often associated with a chaotic life-style and alternating periods of intoxication and withdrawal with rapidly fluctuating blood levels.

If the mother is physically dependent on psychoactive drugs at the time of delivery, the neonate is likely to suffer from an abstinence syndrome because the blood-borne supply of drugs on which the baby has also become dependent is abruptly cut off at birth. Babies born to mothers dependent on sedative-hypnotic such as barbiturates or benzodiazepines may be hyperactive, irritable and restless and may sometimes suffer convulsions. Because of the long duration of action of many of these drugs, the onset of symptoms may be delayed sometimes for as long as 4-7 days after birth. This delay may be hazardous as mother and baby may be discharged before the withdrawal syndrome becomes apparent especially if the mother's dependent state was not diagnosed.

The disadvantageous start to life of many of these babies may be perpetuated if maternal drug use persists after birth when the child's physical, emotional and behavioural development may be further affected by growing up in a drug-taking environment. Finally, one should note that the increased risk of AIDS with which drug abuse is associated, extends to the children of drug abusers because there is a 25%-40% risk of the virus being transmitted to the fetus.

This paper has concentrated on four areas of problems associated with the use and abuse of psychoactive drugs - drug overdose, drug dependence, AIDS and the effect of maternal drug abuse. It is by no means a comprehensive list of all possible adverse consequences. Consideration of these four broad areas, however, shows ample reason why all of us should be concerned about these drugs and why we should all work energetically to ensure that they are used rationally at all times.

## References

1. AA.VV. . DHSS Statistics and Research Division, HMSO, London, 1992.
2. Ghodse A H, Khan I. Psychoactive drugs: improving prescribing practices, WHO, Geneva, 1988.
3. INCB . Narcotic drugs: Estimated World Requirement for 1993; Statistics for 1991, United Nations, Vienna, 1992.
4. Ghodse A H. Drug problems dealt with by 62 London Casualty Departments. Brit J Prev & Soc Med 30:251-256, 1976.

# 12
# The Street Life and Treatment Response of 105 Heroin Addicted Women

**L. M. Gunne, L. Grönbladh, L. S. Öhlund**
Department of Psychiatry, University of Uppsala, S-750 17 Uppsala, Sweden

**Summary**

The street life and subsequent methadone maintenance treatment of 105 female heroin addicts is described and contrasted against the corresponding phases of 244 male heroin addicts. Seventy percent of the females were prostitutes, while the rest relied on criminal activity for their heroin supply. Half of the women were used as sources of heroin money by their male partners and, as expected, this was more common among the prostitutes than the non-prostitute women. Only in 18% the male partners contributed to the illegal heroin money of the prostitutes. Money for food and housing were in all instances obtained from the social welfare system. The sociopsychiatric background for the women showed more heredity for mental disturbance, more broken homes and school problems than the men and the prostitutes had an earlier drug debut. Despite their sinister background the women had a lower mortality rate (47 times the expected) compared with the men (70 times the expected).

Acceptance into methadone treatment (MT) brought about major changes in the life of a majority of the patients. Most of them left criminal activity and prostitution and took up work and studies. However, in about 30% there was an overconsumption of benzodiazepines or alcohol, with a tendency to increase among HIV positive individuals when AIDS symptoms began to emerge. During 5 years of follow-up there was also a persisting opioid abuse in 11-13%. Ex-prostitutes had a somewhat higher opioid abuse than female non-prostitutes. Most MT patients ceased to live with a drug-addicted partner and within 5 years around 45% found a non-addicted partner. There were no new cases of HIV infection within the MT program.

In most treatment institutions for drug addicts women are a minority (generally about 1/3) and the literature on drug addiction thus mainly deals with the drug addiction of men. In the present study we have instead tried to focus on 105 female heroin addicts in a methadone treatment (MT) program. For comparison we present 244 male heroin addicts from the same program. We have also studied mortality rates among 115 street heroin addicts (45 female, 70 male), a majority of whom were on our waiting list for at least 5 years while the intake into our National MT program was closed (1). This group of untreated controls could only be studied with respect to mortality.

All patients presented in this study were selected for MT using acceptance criteria presented earlier (2), and most of them received an inpatient methadone induction in our unit at Uppsala. However, on 1 January 1988 the Swedish Board of Health and Welfare decided that our National MT program, which had been in operation since 1966, should be subdivided in three. This subdivision was undertaken at a time-point when we had 167 patients. When treatment was regionalized in 1988, Uppsala received only 19 patients, whereas 148 were transferred to other programmes. These changes could not be executed immediately and between 1988 and 1990 most Swedish MT patients were still admitted via our treatment group at Uppsala, treatment principles remained the same as before and all patients admitted were followed-up by us. During the last few years, however, which are not covered by the present study, the MT program in Stockholm has undergone great changes and has taken over all aspects of treatment, including the screening procedure for admission.

## 1. Statistical Methods

Survival curves for groups of street heroin addicted men and women were estimated by the product limit method of Kaplan & Meier (3). The log-rank test was used to study differences in survival between the groups. When data were continuous, t tests were performed. In case of unequal variances, the approximate t was computed and Satterhwaite s approximation for the degrees of freedom was used. For categorical data, following Siegel s (4) recommendations, chi-square tests with Yates correction for 4-fold tables were used. The McNemar test for the significance of changes were calculated according to Siegel (4) when the treatment effects were tested. All statistical tests were two-tailed. They were, with the exception of the McNemar test, performed using the statistical package supplied by SAS Institute (5, 6).

## 2. The Street Life of Heroin Addict

### *2.1 Mortality Studies in a Cohort of Untreated Heroin Addicts*

We have earlier shown (1) that a group of untreated street addicts (45 women, 70 men) had a mortality rate that was 63 times the expected, when compared with official Swedish statistics (7). Figure 1 provides a plot of % surviving against age and shows that women had a higher survival rate than men. A Kaplan-Meier analysis (3) gave a highly significant sex difference (p 0.001). The yearly death rate for male street heroin addicts was 9.11% ( which is 70 times the expected for age-matched male Swedish citizens) and 4.21% for female heroin addicts (47 times the expected). Thus the gender-related differences in survival, which exist also in the normal age-matched Swedish population, are accentuated among heroin addicts.

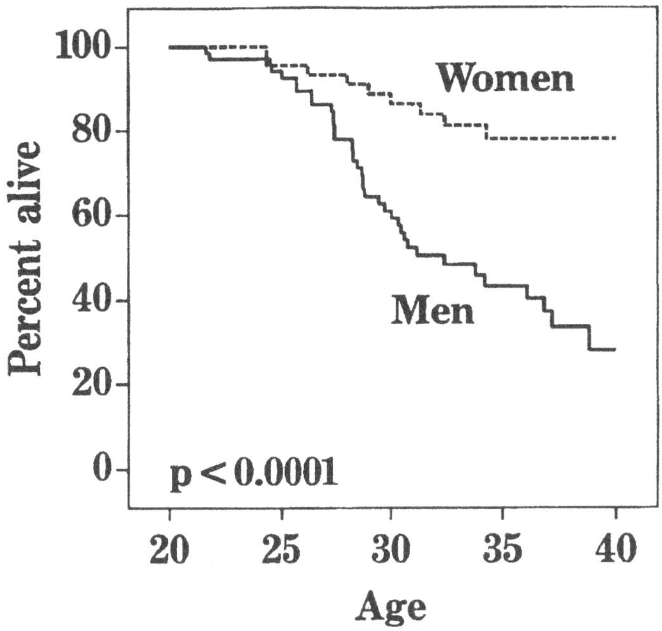


**Fig. 1.** Survivals of 45 female and 70 male street heroin abusers during 5 to 8 years observation

*2.2 Background Information Obtained from MT Patients*

Information concerning demographic data, drug abuse, criminality and sociopsychiatric data was obtained from medical records based on standardized forms, which have been developed for research purposes (8). The information was corroborated by interviews with relatives and friends and by comparisons with earlier hospital records and reports from social agencies and correctional institutions.

Table 1 shows that prostitutes differed both from men and non-prostitute women with regard to numerous drug abuse-related circumstances. The prostitutes had a lower age at drug debut and heroin debut. They were earlier subjected to hospital detoxification and they had more treatment periods (short and long-term) for drug abuse, both in hospitals and other treatment institutions. On the other hand, male heroin addicts spent more time in prison.

Table 2 shows some background factors where the whole group of women differed significantly from men. The women more often had heredity for mental disturbance, including drug or alcohol abuse. They came more often from broken homes, had more school problems, including truancy, and they received more hospital treatment for overdoses. It was noticed that the women had stitch marks and reported injections into the jugular veins more often than the men, revealing a desperate injection technique when arm veins have been used up.

**Table 1.** Sex distribution and percentage prostitute vs non-prostitute females with regard to certain background variables

| | Male (n=243) | Female (n=102) | Analysis | Non prostitutes (n=31) | Prostitutes (n=71) | Analysis |
|---|---|---|---|---|---|---|
| Age at drug debut | | | | | | |
| Mean | 15.8 | 15.2 | t=1.45 | 16.7 | 14.5 | t=2.3 |
| SD | 4.0 | 3.6 | df=343 | 4.6 | 2.8 | df=40 |
| Range | 8-39 | 9-29 | ns | 11-29 | 9-21 | p<0.03 |
| Age at heroin debut | | | | | | |
| Mean | 19.1 | 19.1 | t=0.15 | 20.5 | 18.4 | t=2.1 |
| SD | 3.8 | 4.1 | df=343 | 5.0 | 3.6 | df=44 |
| Range | 12-39 | 12-33 | ns | 13-33 | 12-31 | p<0.05 |
| Age at first detox in hospital | | | | | | |
| Mean | 23.3 | 22.9 | t=0.70 | 25.9 | 21.5 | t=2.71 |
| SD | 5.4 | 6.1 | df=342 | 8.5 | 4.0 | df=36 |
| Range | 1-42 | 14-53 | ns | 16-53 | 14-34 | p<0.02 |
| Hospital periods <7 | | | | | | |
| Mean | 6.5 | 10.8 | t=3.94 | 9.1 | 11.5 | t=1.11 |
| SD | 7.1 | 10.1 | df=144.9 | 10.1 | 10.0 | df=100 |
| Range | 0-60 | 0-50 | p<0.001 | 1-50 | 0-45 | ns |

**Table 1.** Sex distribution and percentage prostitute vs non-prostitute females with regard to certain background variables (continued)

| | Male (n=243) | Female (n=102) | Analysis | Non prostitutes (n=31) | Prostitutes (n=71) | Analysis |
|---|---|---|---|---|---|---|
| Hospital periods≥7 | | | | | | |
| Mean | 4.0 | 5.3 | t=2.80 | 4.2 | 5.8 | t=1.71 |
| SD | 3.4 | 4.4 | df=152.1 | 4.1 | 4.5 | df=100 |
| Range | 0-25 | 0-20 | p<0.007 | 0-20 | 0-20 | ns |
| Month at therapeutic community | | | | | | |
| Mean | 4.5 | 8.1 | t=2.32 | 7.8 | 8.2 | t=0.11 |
| SD | 8.7 | 14.4 | df=133.2 | 18.4 | 12.4 | df=42 |
| Range | 0-79 | 0-99 | p<0.03 | 0-99 | 0-72 | ns |
| Months in prison | | | | | | |
| Mean | 21.1 | 6.9 | t=5.75 | 7.5 | 6.7 | t=0.24 |
| SD | 31.2 | 14.4 | df=340.1 | 14.5 | 14.5 | df=100 |
| Range | 0-154 | 0-68 | p<0.0001 | 0-60 | 0-66 | ns |
| Age at entering MT | | | | | | |
| Mean | 30.0 | 29.5 | t=0.62 | 31.2 | 28.8 | t=2.04 |
| SD | 5.6 | 5.4 | df=343 | 5.7 | 5.1 | df=100 |
| Range | 20-53 | 20-46 | ns | 23-46 | 20-43 | p<0.05 |

**Table 2.** Socio-psychiatric background facιors (%) yielding significant gender-related differences (Males N=244, Females N=102)

| | Males | Females |
|---|---|---|
| Heredity for mental disturbance | 32 | 45 |
| Broken home | 49 | 62 |
| Truancy at school | 6 | 29 |
| Frequent overdosess | 30 | 45 |

Our data seem to indicate that female heroin addicts have a more severe social handicap and since they are probably fewer in numbers, they may represent a more socially and psychologically selected group than male heroin addicts. Therefore their greater ability to survive is remarkable.

*2.3 The Street Life of Heroin Addicted Prostitutes*

Seventy percent of the women in this study were street prostitutes and a majority of these 74 prostitutes were used as sources of money for heroin by their addicted partners. There did not seem to be a system of pimps, with more than one girl, but the "boyfriends" were exchangeable and whenever one of them went to prison, his place was taken by another heroin addict who exploited the woman.

A regular heroin addicted street prostitute has about 1500 sexual contacts per year and makes around $ 100,000 per year (Table 4). Some of these women have fewer contacts and have built up a network of regular customers, who have to pay more. The average yearly taxfree income of the prostitutes was $ 132,000 and these 74 women thus delivered around $ 27,000 per day to the illicit heroin market. They all pay for food, housing and other everyday expenditures by money from the social welfare bureau. In addition most of them are shoplifters for clothes, makeup, etc. and they are often sentenced for petty crimes. The year before entering MT 79% had been tried for such criminality.

From an economic point of view the heroin addicted men live as parasites on their prostitute girl friends. Two thirds of the street prostitutes finance both their own and their male partners' heroin. Prostitution and heroin addiction are mutually reinforcing behaviors. These patients consider prostitution to be the only quick way to obtain money for heroin and without heroin they wouldn t stand the life of a prostitute.

### *2.4 Non-Prostitute Heroin Addicted Women*

Thirty-one of the 105 heroin addicted women were not prostitutes, but preferred to maintain a regular supply of heroin through criminal activity. They were regularly sentenced for serious crimes (burglary, robbery, theft, etc). The year before entering MT 68% of the non-prostitutes had been tried at court for various crimes (the difference between prostitutes/non-prostitutes not significant).

While 61% of the prostitutes finance the heroin of their male partners, only 29% of the non-prostitutes do so. Non-prostitutes seem to enjoy more equality in relation to their male heroin addicted partners. In 55% both partners contribute to heroin money, while in the households of prostitutes the corresponding figure is only 18%.

## 3. Efficacy of the MT Program

### *3.1 Treatment Strategy*

Our treatment philosophy favors a rapid return to working life, facilitating the acquirement of new friends and regular habits. Only a minority in our experience seem to need or benefit from individual or group psychotherapy. Instead such measures may pacify patients, while postponing their inevitable confrontation with realities to a time point when their own expectations and enthusiasm for the new treatment may be lower.

Most patients were started as inpatients and were initially detoxified. After negotiations with the patient, a rehabilitation plan was agreed upon and methadone induction was started. After a while it was found that newly admitted patients knew what was expected and had a rehabilitation plan ready soon after arrival. The methadone dose was often adjusted by aid of plasma levels (9). The mean dose was 80 mg/d, both in men and women (range 30-130 mg/d).

The development of our patients, particularly during the early phase of treatment, was closely watched by the treatment staff and control and surveillance measures adjusted accordingly. When there were signs of resocialization, they received strong support and encouragement and at the same time suspicious and integrity-infringing control measures were reduced. A newly-admitted patient, who is used to being pushed around in the treatment system, may not take offense if he is subjected to urine sampling under eye control. But later on, when he is climbing the social ladder, he has to be treated differently. For research purposes all patients delivered 2-3 urine samples/week throughout the study, but only about 10% were delivered in the presence of a nurse.

### *3.2 Effects on Criminal Activity*

The year before entering MT 82% of the men and 76% of the women had been

**Table 3.** Evaluation of a methadone treatment program. Effects on criminality (per cent subjects sentenced for low violations/year). Year 0 means the year before entering treatment. The McNemar Test compares each treatment year with year 0

| | Year | | | | | |
|---|---|---|---|---|---|---|
| | 0 | 1 | 2 | 3 | 4 | 5 |
| Men | 81.5 (198/243) | 9.2 (22/240) | 8.7 (17/196) | 9.5 (14/148) | 12.2 (14/115) | 12.0 (12/100) |
| All women | 75.5 (77/102) | 7.0 (7/100) | 2.5 (2/79) | 0.0 (0/58) | 6.8 (3/44) | 7.9 (3/38) |
| McNemar test | | $\chi^2$=237.1 p<.001 | $\chi^2$=198.4 p<.001 | $\chi^2$=137.5 p<.001 | $\chi^2$=88.5 p<.001 | $\chi^2$=72.7 p<.001 |
| Prostitutes | 78.9 (56/71) | 8.6 (6/70) | 1.8 (1/55) | 0.0 (0/41) | 6.3 (2/32) | 7.4 (2/27) |
| Non-prostitutes | 67.7 (21/31) | 3.3 (1/30) | 4.2 (1/24) | 0.0 (0/17) | 8.3 (1/12) | 9.1 (1/11) |
| McNemar test | | $\chi^2$=66.0 p<.001 | $\chi^2$=57.0 p<.001 | $\chi^2$=43.0 p<.001 | $\chi^2$=32.0 p<.001 | $\chi^2$=23.3 p<.001 |

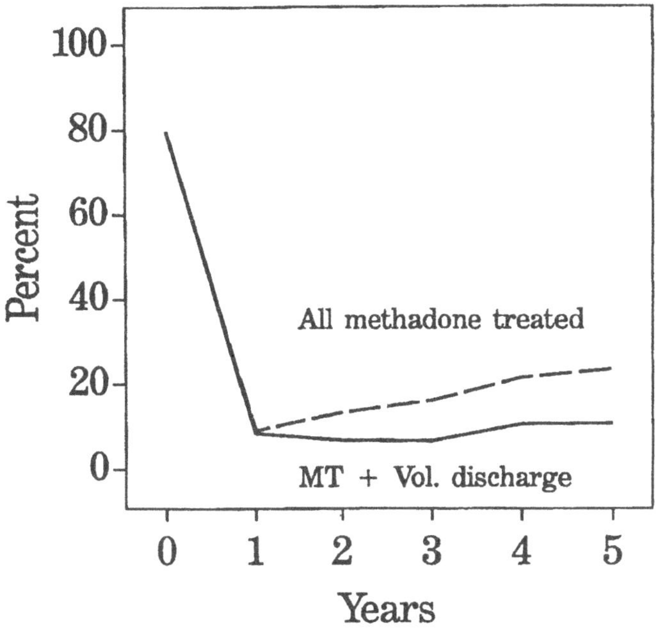


**Fig. 2.** Percent patients with criminal activity resulting in court trials during 5 years in an MT program. Year 0 is the years before admittance. Solid line: patients in treatment. Broken line: all patients

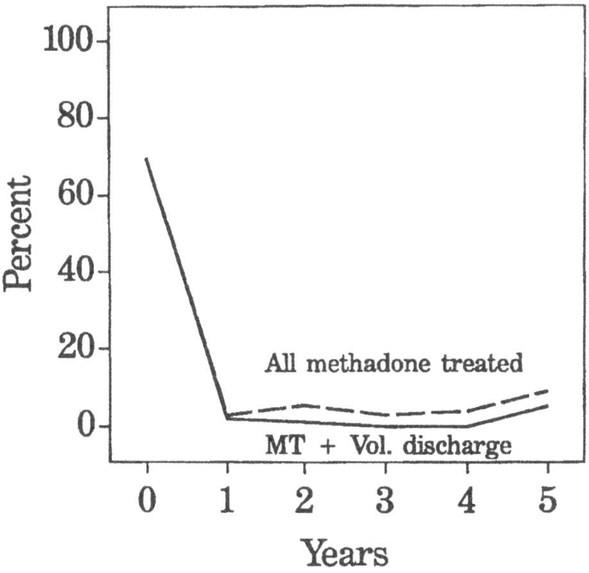


**Fig. 3.** Percent prostitution among females before and during 5 years in MT. Solid line: prostitution in treatment. Broken line: all female patients

**Table 4.** Sexual contact/year among heroin addicted street prostitutes

| | N | Contacts/year | Income/women |
|---|---|---|---|
| Street prostitutes | 49 | 1 535l418 | $109 642 |
| Regular customers | 25 | 1 241l321 | $177 285 |
| Mean | | | $132 494 |

Before MT the whole group of 74 prostitutes were delivering $ 26 862/day to the illicit heroin market. After 5 years this figure had dropped to $ 1 815/day

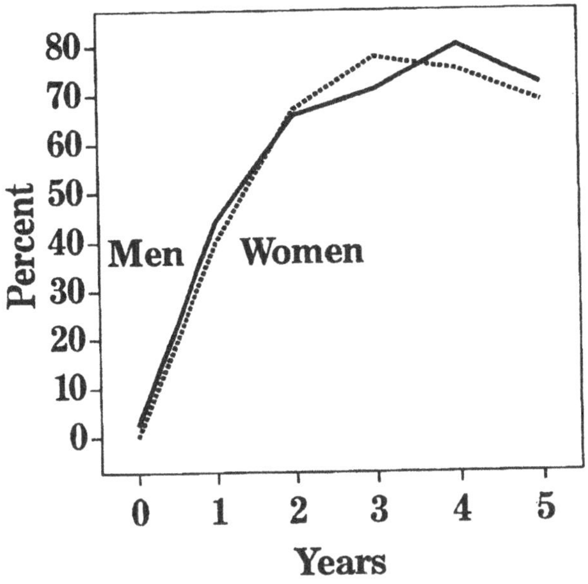


**Fig. 4.** Work and studies among men and women during 5 years in MT. The years before acceptance, year 0, there were 2.5% men and 0% women working

sentenced for law violations. During the 5 years of follow-up this yearly percentage varied between 8.7 and 12.2% for the men and 0-7.9% among the women (Table 3). There were no significant differences between prostitutes and non-prostitutes in this respect during years 1-5.

Since certain criminal activities are regarded as dischargeable offense, and since imprisonment means at least temporary exclusion from MT, Fig.2 illustrates the effect on criminal activity of all who were accepted for treatment (broken line), as well as those in treatment or voluntarily discharged with medical consent (solid line). The treatment effect is still obvious and statistical significant, when rates of criminality are compared according to McNemar test before (year 0) and each year after acceptance (years 1-5).

### *3.3 Effects on Female Prostitution*

For several years we had a close cooperation with 2 groups in Stockholm (one from the police, the other from the social welfare bureau) who studied the development of prostitution. They knew all the girls and who had been accepted for MT and they immediately alerted us if one of "our prostitutes" had been seen in the vicinity of the prostitution areas.

Figure 3 shows that there were marked and significant effects on prostitution, which was 70% before treatment (year 0), but nearly disappeared during the 5 follow-up years. Relapse into prostitution behavior mainly occurred among the few who were excluded from treatment due to polydrug abuse. They are included among "all methadone treated" in Fig.3 (broken line). The slight increase year 5 was due to one prostitute and one non-prostitute who took up street prostitution (and were later expelled).

### *3.4 Work and Studies*

Both men and women began to work or study, thus returning to a regular social life in 70-80% of the cases (Fig.4). According to McNemar test this increase was highly significant. Around 30% of those jobs were relief work, due to the increasing difficulties to obtain regular employment in times of economic depression. Since our treatment program puts great emphasis on vocational rehabilitation, all patients are encouraged to seek jobs and to prove that they are still working by sending us their monthly paycheck stubs. This is a precondition for receiving take-home privileges and lessens the need for intense urinalysis surveillance. With regard to work, prostitutes responded at least as well as non-prostitutes.

### *3.5 Heterosexual Partners*

An important sign of success in a drug abuse treatment program is reflected by the patients' relations to their partners. A non-abusing person can hardly live together with an actively abusing drug addict. For this reason we studied what would happen to the partnerships of our patients after acceptance in an MT program (as therapists we are not

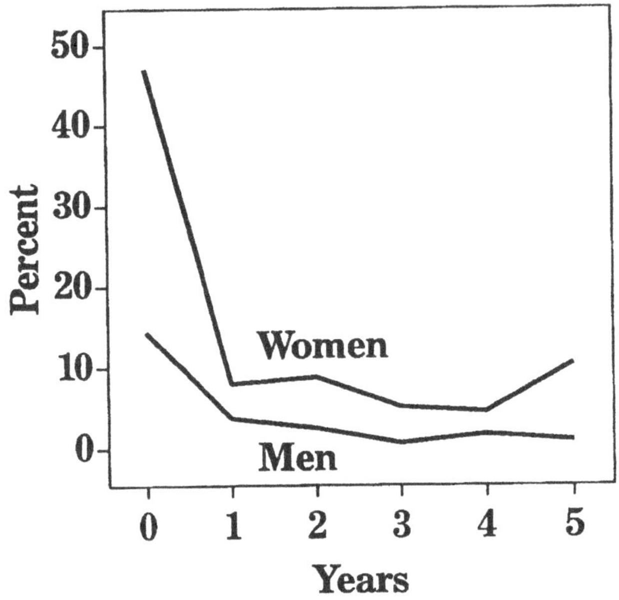


**Fig. 5a.** Patients living together with drug-addicted partner

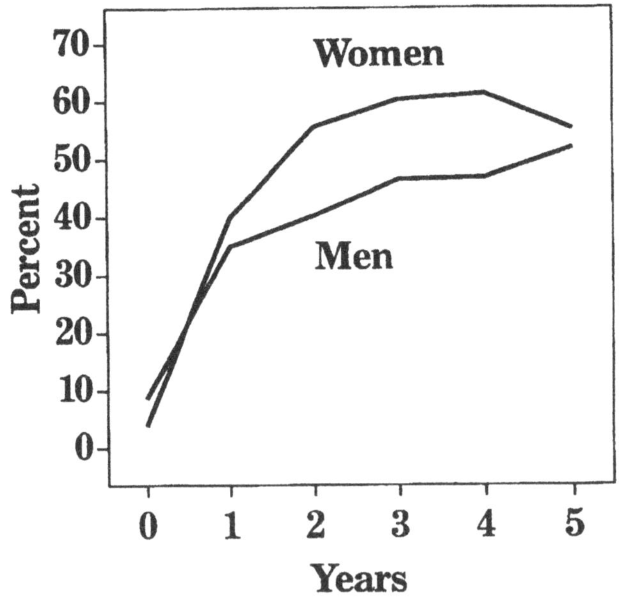


**Fig. 5b.** Patients living together with drug-free partner

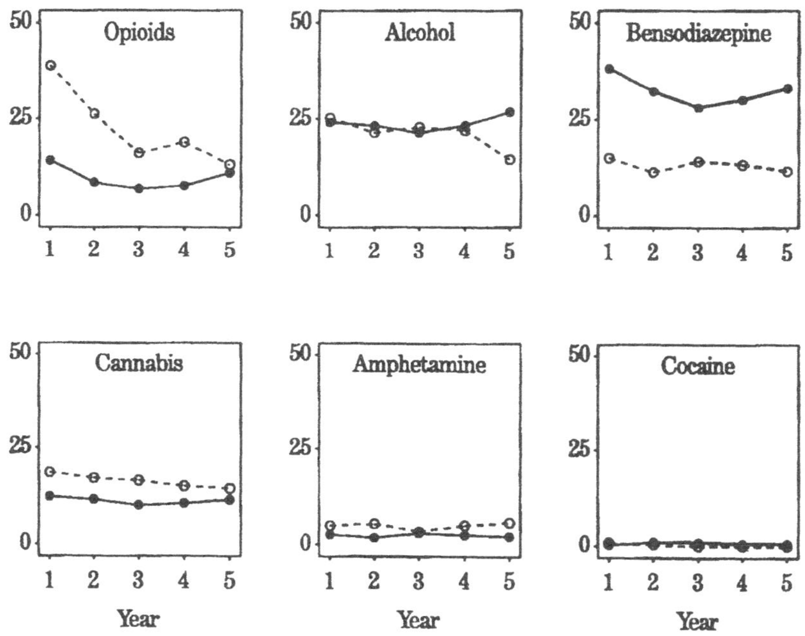


**Fig. 6.** Solid lines: percent individuals with abuse (>once/month) of drugs and/or alcohol during 5 years in MT. Broken lines: occasional use (<once/month)

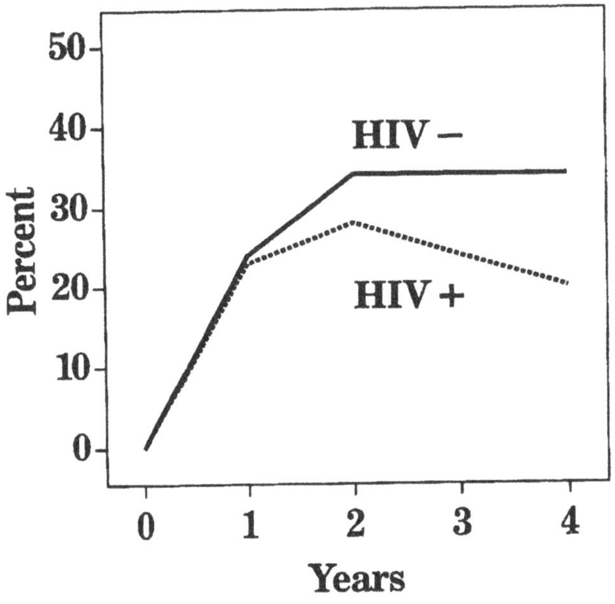


**Fig .7.** Percent patients entirely free of drug or alcohol abuse during 5 years in MT. Solid line: HIV negative (n=266), broken line: HIV positive (n=74)

trying to interfere with this development, but are merely recording spontaneous changes). During the first 3 years after acceptance, drug addicted partners were abandoned by our patients, men and women alike (Fig.5a), and after some time MT patients found drug-free partners (Fig.5b). The percentage men and women living alone showed a moderate decline during the 5 observation years. The recorded change of partners seems to prove that our patients have undergone a thorough alteration in life style.

*3.6 Drug and Alcohol Abuse*

Throughout the study urine samples were monitored 2-3 times a week. Initial screening was made by the EMIT technique, with tests for methadone, morphine-like opioids, benzodiazepines, barbiturates, amphetamines, cocaine (ekgonine metabolites), and cannabis. Positive urines were further tested using GC/MS and ethanol was monitored by gas chromatography.

While treatment was successful in many important respects, a high consumption of drugs remained a problem that lingered for many years and which may not always have been solved in an optimal way. Partly this may be explained by our patients earlier polydrug abuse (though severe dependence on non-opioid drugs is among the exclusion criteria when patients are selected for MT).

There were 3 distinguishable patterns of drug and alcohol use in the program. One third became altogether free of drug or alcohol abuse. Around 40% initially had an occasional use of opioids (no more than once per month), which dropped down to 10%. Finally there was a group (about 30%) with consistent, and in a few instances even increasing use of benzodiazepines and/or alcohol. In this group the use of opioids and/or cannabis did not disappear entirely during the 5 years of observation. Prostitutes were more often among the opioid abusers during the first year than non-prostitute women ($p<0.05$). The yearly percentage of prostitutes with opioid abuse during the 5 treatment years was 17.1; 16.4; 14.6; 9.4 and 11.1%, while the corresponding figures for non-prostitutes were 3.3; 4.7; 5.9; 8.3 and 0%. This may indicate a more severe dependence on opioids among the ex-prostitutes, who may have had easier access to heroin in the past. Amphetamine (which is a common drug of debut among Swedish heroin addicts) and cocaine remained low throughout the study (Fig.6). There were no significant differences between male and female patients with regard to drug abuse in the program.

### *3.7 Effects on HIV Infection*

In 1984, when HIV and AIDS had become a reality, and our treatment program was reopened for new admissions after 5 years of closure, we tested all 67 who were then in treatment. Two were found to be HIV positive. They were two female ex-prostitutes living together with HIV infected men (former regular customers). These women may have acquired the contagion via sex, not syringes. Since that time 40% of the newly admitted patients were found to be HIV positive, but no one seems to have acquired the virus after admittance (10). This further confirms the change in life style among our patients.

There was a tendency among HIV positive to become less drug-free with time (Fig.7). They gradually took up use of benzodiazepines when AIDS symptoms became increasingly prominent.

## 4. Discussion

Available data seem to indicate that female heroin addicts represent a selection with more severe sociopsychiatric handicaps compared with male heroin addicts. Among the women, prostitutes showed signs of a more intensive heroin use before MT than non-prostitute women (earlier drug debut, more detox and other treatments) in accordance with earlier reports (11). The lives and habits of heroin abusing female prostitutes appears to involve great personal risks, including a proliferous access to heroin. A stronger dependence on heroin may be the reason why prostitutes had more relapses into drug-taking during treatment than non-prostitutes. In spite of their intense involvement with heroin the death rate of female street heroin addicts was only half that of male addicts (4.2% vs 9.1% deaths per year).

We have presented outcome data from a cohort of patients who were followed for 5 years, which is not a common method of evaluation. Our results indicated major beneficial effects on reduction of crime and prostitution and improvement with regard to vocational rehabilitation. The effects on drug and alcohol abuse were less outstanding, but a comparison with published results from other programmes reveals problems of this kind in most treatment programs.

Ball and Ross (12) in a survey of 6 US treatment programmes reported very good effects on crime reduction. There was a 79% reduction of criminal activity after 6 months or more in treatment and the number of crime days were reduced by 91%. Groups of patients who were newly admitted or had been 1 year, 2 years, etc. in treatment were compared, but no prospective follow-up was undertaken. Sechrest and Crim (13) likewise found crime reduction on follow-up 2 and 3 years after admittance in two different MT programmes. Since the pretreatment level of criminality was expressed as total arrests during the whole period of addiction, the results are not readily comparable with ours.

With regard to drug abuse during treatment our patients had an 88% reduction of i.v. heroin use, while the combined US studies reported a 71% reduction in i.v. use (12). The American studies showed a great variation in current i.v. drug use from 10 to 57% between the different programmes (12). Hser et al. (14) reported that after 2 months in a Californian MT program most patients had reverted to daily heroin use. Non-opioid use was rather high in the American programmes (cocaine 18-19%, cannabis 37-42%, sedatives 18-19%, alcohol intoxication 30%). A high rate of cocaine use (71%) was also reported from a New York City program (15). Probably local drug habits and preferences may influence the outcome of MT. We had expected to find more amphetamine use in a Swedish program, since there is an endemic amphetamine use in Sweden since the sixties and many of our patients had earlier been amphetamine users. However, no such problems have emerged.

The early American studies reported on results very similar to our own (16). There was a marked decrease in antisocial behavior and increase in social productivity in up to 85% after 5-6 years in treatment. These good results proved to be difficult to maintain as MT grew to vast dimensions in the US. In the Swedish program, however, with its limited size, results have remained satisfactory for 25 years. The beneficial effect of MT on risk reduction with regard to HIV has been confirmed in other studies (17). The finding that no patient has become HIV positive in our program since 1985 supports the notion that our patients have undergone a major change in life style. This is further confirmed by the observation that they abandon drug abusing partners and seek new, drug-free relations.

## References

1. Grönbladh L, Öhlund L S, Gunne L M. Mortality in heroin addiction: impact of methadone treatment. Acta Psychiatr Scand 82:223-227, 1990.
2. Grönbladh L, Gunne L M. Methadone-assisted rehabilitation of Swedish heroin addicts. Drug Alc Dep 24:31-37, 1989.

3. Kaplan E L, Meier P. Non-parametric estimation from incomplete samples. J Am Statist Assoc 53:457-481, 1958.
4. Siegel S. Non-parametric statistics, Mc Graw-Hill, New York, 1956.
5. SAS Institute. SAS/STAT Guide for personal computers, version 6, SAS Institute, Cary, NC, 1987.
6. SAS Institute. Technical report P-179. Additional SAS/STAT procedures release 6.03, SAS Institute, Cary, NC, 1988.
7. AA.VV. . Statistical Abstracts of Sweden 1989. Official Statistics of Sweden, Norstedts Tryckeri, Stockholm, 1988.
8. Frykholm B, Gunne L M, Huitfeldt B. Prediction of outcome in drug addiction. Addictive Behav 1:103-110, 1976.
9. Holmstrand J, Änggård E, Gunne L M. Methadone maintenance: plasma levels and therapeutic outcome. Clin Pharmacol Ther 23:175-180, 1978.
10. Blix O, Grönbladh L. The impact of methadone maintenance treatment on the spread of HIV among IV heroin addicts in Sweden. In: Drug Addiction and AIDS, edited by Loimer N, Schmid R. Springer Verlag, Wien, New York, 1991, pp 200-205.
11. Kuhns J B, Heide K M, Silverman I. Substance use/misuse among female prostitutes and female arrestees. Int J Addict 27:1283-1292, 1992.
12. Ball J C, Ross A. The effectiveness of methadone maintenance treatment, Springer Verlag, New York, 1991.
13. Sechrest D K, Crim D. Methadone programs and crime reduction: a comparison of New York and California addicts. Int J Addict 14:377-400, 1979.
14. Hser Y, Anglin M D, Liu Y. A survival analysis of gender and ethnic differences in responsiveness to methadone maintenance treatment. Int J Addict 25:1295-1315, 1991.
15. Kang S Y, De Leon G. Criminal involvement of cocaine users in a methadone treatment program. Addiction 88:395-404, 1993.
16. Gearing F R, Schweitzer M D. An epidemiologic evaluation of long-term methadone maintenance treatment for heroin addiction. Am J Epidemiol 100:101-112, 1974.
17. Bellis D J. Reduction of AIDS risk among 41 addicted female street prostitutes: effects of free methadone treatment. J Addict Dis 12:7-23, 1993.

# 13

# The Pharmacological Rationale for Methadone Treatment of Narcotic Addiction

**R. G. Newman**
Beth Israel Medical Center, New York City, NY, U.S.A.

## Summary

Despite over a quarter-century of remarkably consistent reports from throughout the world of favorable results with the use of methadone to treat heroin addiction, this modality remains exceedingly controversial. Even among those who utilize methadone in caring for addicts, there is intense debate over optimal dosages, duration of treatment and other aspects of the medical care which is rendered. Ironically, the polarization seems to reflect a misunderstanding of the basis pharmacological rationale for methadone treatment.

In this paper the pharmacological principles which apply to narcotics in general, and to methadone in particular, are presented. It will be noted that the effects (and, in maintained patients, the lack of effect with respect to euphoria) of methadone are absolutely free of any controversy. Physicians and policy-makers must be guided by these straight-forward, unequivocal pharmacological properties in determining how methadone should be utilized in the management of patients with a chronic, potentially deadly illness.

## Pharmacological Rationale for Methadone Treatment

I can not emphasize too strongly the fact that the pharmacology of methadone is not and never has been the subject of debate among knowledgeable professionals. There are moral issues, political issues, economic issues and pragmatic issues which can provoke respectful disagreement and thoughtful discussion, but the way methadone affects (and does not affect) the body is absolutely unequivocal.

Methadone is a narcotic medication, which merely means that it has a series of actions similar to those of morphine. Table 1 provides a partial listing of these actions. There are obviously many drugs which produce one or more of the individual effects listed, but unless they are capable of eliciting the entire spectrum, they are not classified as narcotics.

**Table 1.** Narcotic: a drug which has a series of actions similar to those of morphine

| |
|---|
| CENTRAL NERVOUS SYSTEM DEPRESSION |
| Respiratory depression (death) |
| Sleepiness |
| ANALGESIA |
| EUPHORIA |
| COUGH SUPPRESSION |
| PUPILLARY CONSTRICTION |
| NAUSEA AND VOMITING |
| INCREASED RESPIRATORY TRACT SECRETIONS |
| CONSTIPATION |
| DIAPHORESIS |
| ITCHING |

You will note that nausea and vomiting are among the pharmacological effects of narcotics. And yet, it is obvious if addicts experienced these exceedingly unpleasant consequences daily, each time they inject heroin, we clearly would not be faced with a problem of narcotic addiction! But why does the addict not experience these effects, since they are listed - appropriately - among the pharmacological actions of narcotics? The answer is a simple one: repeated exposure to a narcotic (and indeed, to most substances) produces a tolerance to the drug's effects. Clinically, this is known to physicians and patients alike who experience the great frustration in treating pain, only to find that a tolerance to the analgesic properties of even the most potent narcotics develops quite quickly and soon reaches a stage where no relief is obtainable.

There are several characteristics of tolerance which must be stressed. First of all, tolerance does not develop to the same degree or with the same speed for each of the various actions of the drug. Thus, the addict soon develops tolerance to the adverse effects of nausea and vomiting, but is still able to achieve euphoria, and risks overdose due to the central nervous system depressant effect.

Secondly, although tolerance develops as a result of repeated exposure to a specific drug, it applies to all drugs in the same class. Thus, a patient whose pain has been treated exclusively with morphine, and who develops tolerance as a result, is unable to get analgesic relief from codeine, demerol or any other narcotic. Similarly, although the street addict's exclusive drug may have been heroin, he or she develops a tolerance to morphine, methadone and all other narcotics as well.

And thirdly, although tolerance is defined as a level of drug concentration in the body which must be exceeded before the pharmacological effect can be experienced (as shown in Fig. 1), that level may be so high that it is unreachable regardless of the amount of drug taken. A non-narcotic example of this phenomenon is the common decongestant nose drop. Initially the sufferer experiences great relief, but tolerance quickly develops and to such an extent that even an endless flow of the medicine will produce no effect whatsoever. As mentioned previously, precisely the same phenomenon precludes sustained analgesia through the administration of narcotics, even if they are given in progressively higher and higher dosages.

Along with tolerance, there is another consequence of repeated exposure to narcotics, and that is dependence. Dependence also refers to a level of drug concentration in the body, but it has nothing to do with the pharmacological actions of the drug. Rather, it is the level which must be exceeded in order to avoid the symptoms associated with the absence of the drug. As Fig. 2 indicates, when the concentration of narcotics in the body of a physically dependent individual falls below the dependence level, withdrawal symptoms result, while when the concentration is above this level there are no such symptoms.

As in the case of tolerance, dependence - even when a consequence of exposure to only a single specific drug - is in fact a dependence on the entire class of drugs. This means that withdrawal symptoms in an addict who has never taken anything but heroin can be treated effectively by the administration of any narcotic.

A range exists between the dependence level and the tolerance level which is critical to the understanding of the use of methadone in the treatment of addiction. As indicated in Fig. 3, as long as the concentration of narcotics in the body falls below the tolerance level (thus precluding the patient experiencing any of the narcotic effects), and yet is above the dependence level (thus precluding withdrawal symptoms), the patient will look and feel completely normal. The most astute clinical observer will be unable to distinguish the addict from the non-addict under these circumstances. Of course, the heroin-dependent user, if unable to obtain another fix, will experience withdrawal symptoms in a matter of hours as the drug from previous injections is metabolized and the concentration falls below the dependence level.

Fig. 1.

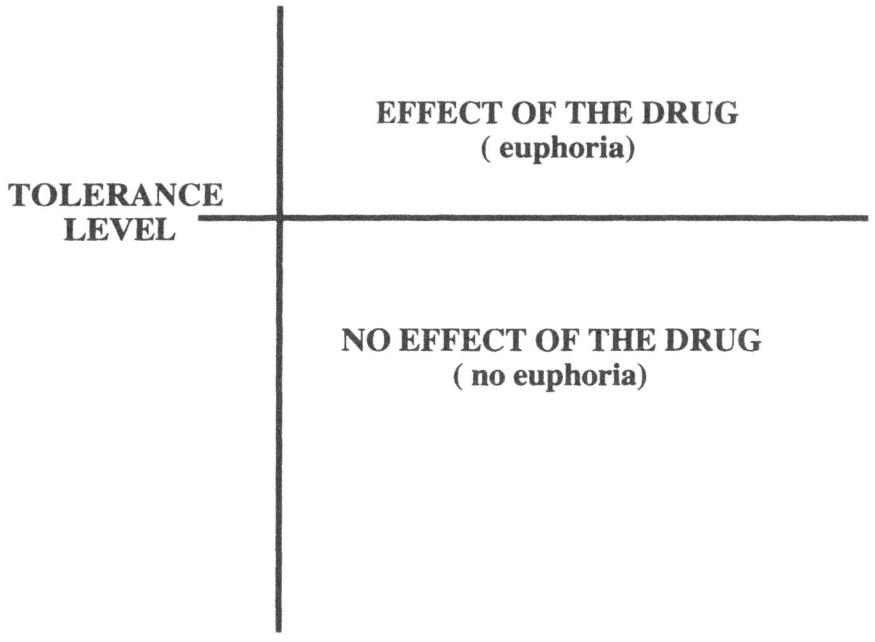

Fig. 2.

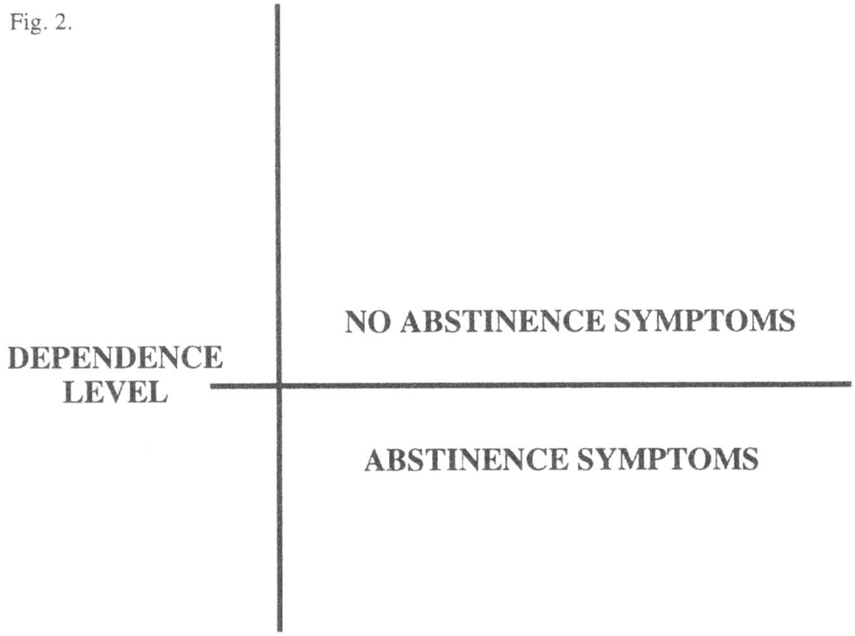
NO ABSTINENCE SYMPTOMS
DEPENDENCE LEVEL
ABSTINENCE SYMPTOMS

Fig. 3.

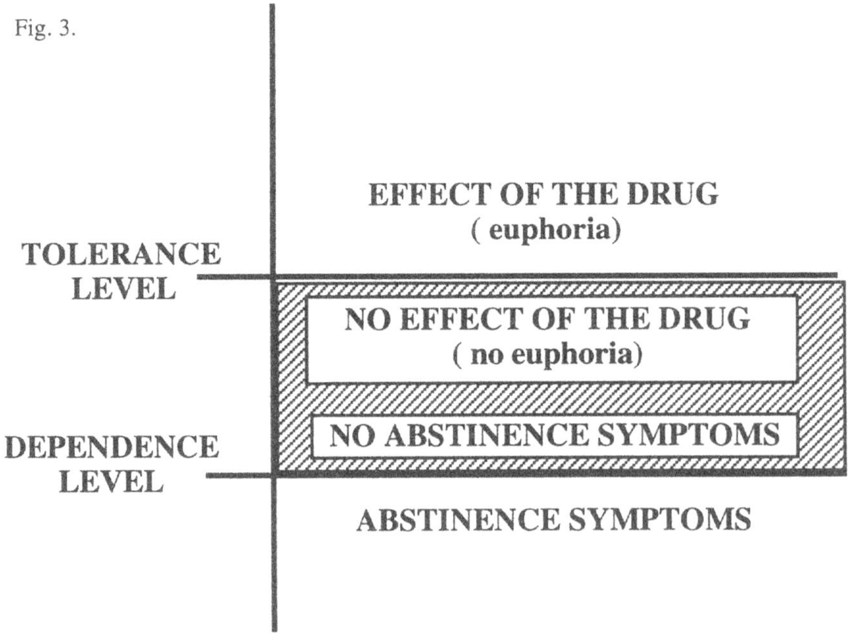
EFFECT OF THE DRUG
( euphoria)
TOLERANCE LEVEL
NO EFFECT OF THE DRUG
( no euphoria)
NO ABSTINENCE SYMPTOMS
DEPENDENCE LEVEL
ABSTINENCE SYMPTOMS

In the chemotherapeutic treatment of addiction, whether the goal is short-term detoxification or long-term maintenance, the objective is obvious: to maintain the patient in a state of physiological normalcy by keeping the narcotic concentration in the body in the range between the tolerance level and the dependence level. Theoretically, one could try to "stabilize" a patient with heroin or morphine, but this would require the administration of the drug at least four or five times each day, and by injection rather than by the oral route (Fig. 4). Obviously, such a treatment regimen would be impossible from a pragmatic standpoint. With methadone, on the other hand, this objective is relatively easy to achieve. Firstly, methadone has a very predictable effectiveness even when taken by mouth; and secondly, the duration of effectiveness is in the neighborhood of 24 to 36 hours, while that of virtually all other narcotics is no more than three to six hours (Fig. 5).

In countries throughout the world where methadone treatment has been employed, it has been demonstrated that an initial dose of methadone of 30 to 40 mgs. will prevent withdrawal symptoms without producing any significant untoward effects. This is true regardless of the amount of heroin being consumed by the individual addict, the purity of the drug or the route of administration. Once treatment has begun, the body quickly adjusts to the starting dosage such that the concentration of the narcotic in the body is maintained at approximately the mid-level between the dependence and the tolerance levels.

Fig. 4.

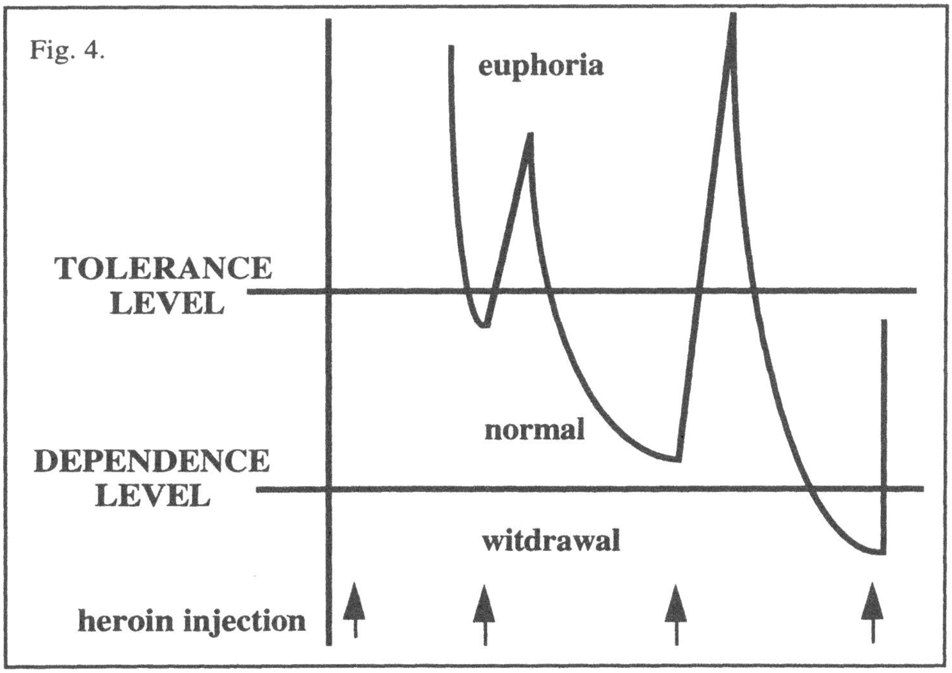

Fig. 5.

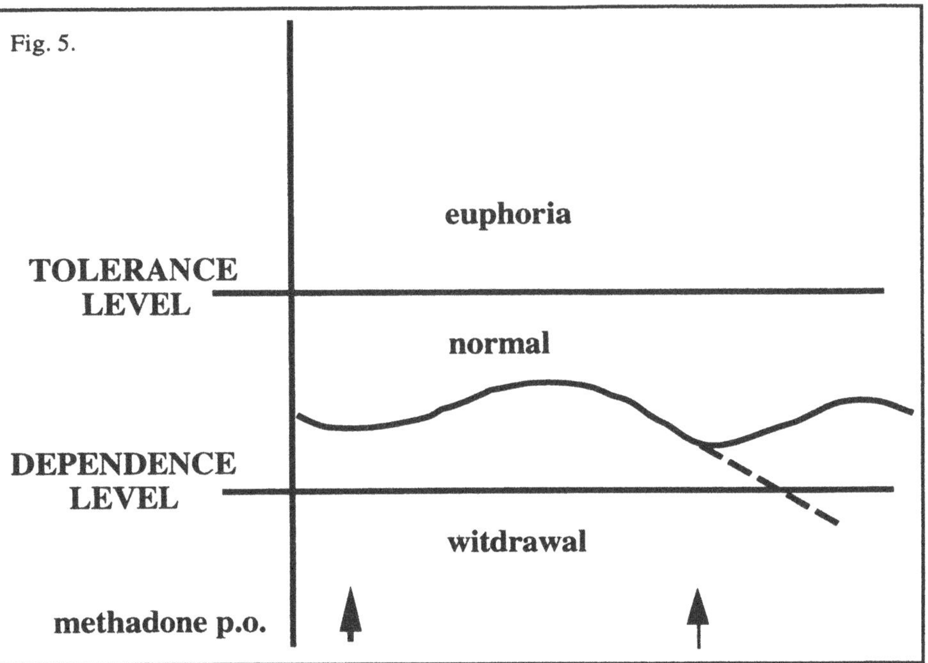


**Methadone Used for Detoxification**

Since we are dealing with a range between the two levels, and a relatively broad range at that, it is possible to increase or decrease the dosage by five or ten milligrams of methadone without crossing either of the two levels. Thus, in detoxification treatment, a decrease of five milligrams will not be accompanied by withdrawal symptoms, and after several days at the new reduced level, the dosage can be lowered once more, and this process can be repeated until methadone administration has been discontinued altogether (Fig. 6). In this way, generally in no more than 14 days, the physical dependence can be eliminated successfully without the patient experiencing withdrawal symptoms.

The benefits of such short-term addiction treatment with methadone are very substantial. It is a completely safe, very effective, relatively inexpensive (particularly when provided on an outpatient basis) medical intervention in the chronic problem of heroin addiction. It lends itself to very rapid and large-scale implementation - to the extent that virtually unlimited numbers of patients can be accommodated promptly. In this way, it can ease significantly the unconscionable situation in which addicts who desperately want treatment must remain on "waiting lists" because long-term rehabilitation programs are filled to capacity. In addition, we know from the experience in many parts of the world that no other form of treatment generates as much demand among the addict population. For example, in New York City, a network of only five ambulatory detoxification clinics admitted over 22,000 individuals yearly in the early 1970s - before the program was terminated by short-sighted government officials determined to reduce expenditures.

On the other hand, there is a major limitation of detoxification treatment: we know that once withdrawal from physical dependence has been accomplished, the former heroin user almost invariably will revert to illicit drug use. This might occur after a matter of days or, in some cases many months, but sooner or later relapse is the rule rather than the exception. For this reason, there is another application of methadone in the treatment of heroin addiction: as a maintenance medication for patients who are motivated to give up heroin use and the life style associated with it.

**Methadone in the "Maintenance" Treatment of Addiction**

Methadone "maintenance" is really a relative term and does not denote any specific duration of treatment. It should be clear from the pharmacological description above that any patient can be maintained - with any amount of methadone - for an indefinite period of time. Constant doses will maintain the concentration of methadone in the range between the dependence and the tolerance levels, and thus neither withdrawal symptoms nor narcotic effects will be experienced. It also should be obvious that any patient can be detoxified from methadone through a gradual reduction of dosage, in precisely the manner described previously, regardless of the dose and duration of methadone treatment.

A fundamental question is: why maintain a patient on methadone? The answer is simple: empirically, it has been demonstrated in countries throughout the world that individuals are able to give up illicit heroin use and to resume (and in many instances

Fig. 6.

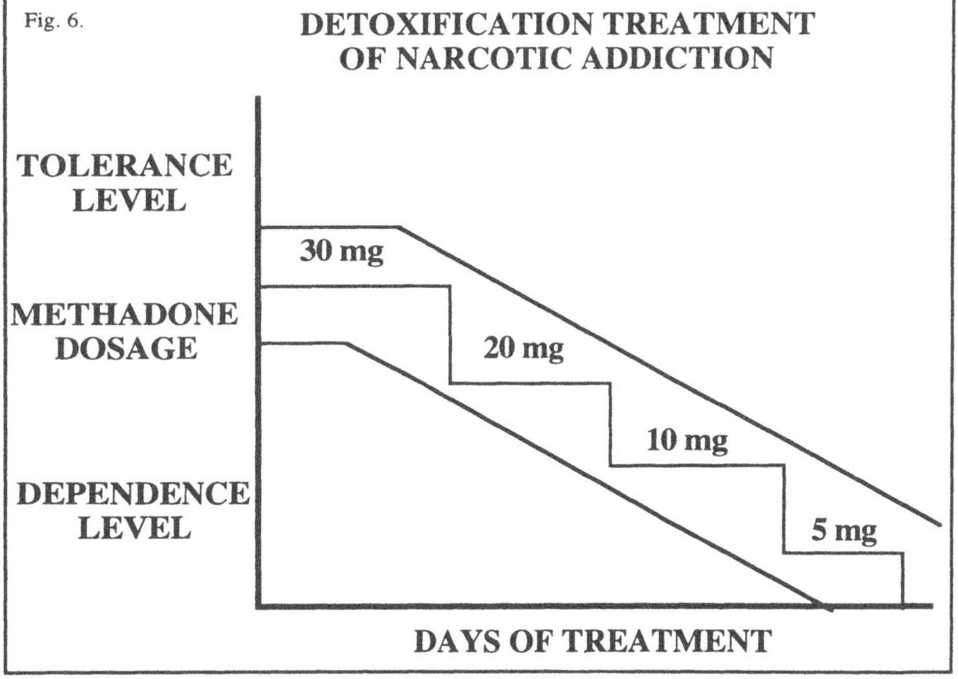

Fig. 7.

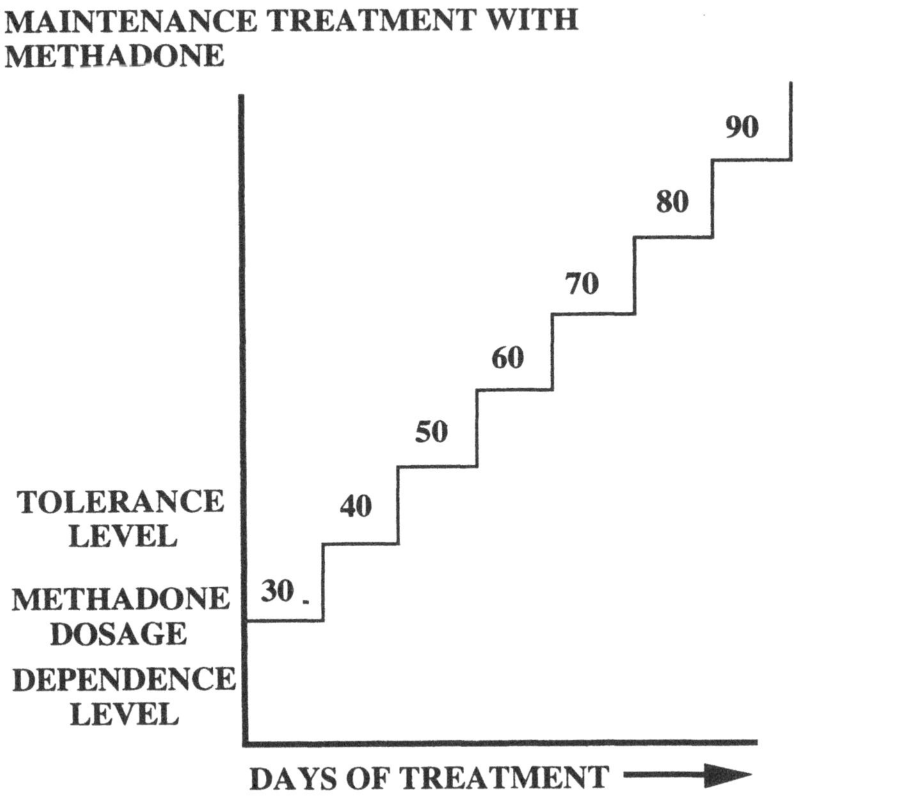


assume for the first time) a normal, productive, socially acceptable and self-fulfilling life style while maintained on constant doses of methadone. Surely this is a compelling rationale for methadone maintenance!

Another question commonly raised is: how long should this treatment continue? The response, again, is both simple and empirical: as long as it is effective! As in any other form of medical treatment, the success of the treatment regimen is determined by how the patient responds, and the favorable assessment of a therapeutic outcome is not diminished in the slightest by the fact that the patient continues to receive medication.

What underlies the response to both of the above questions is the reality that heroin addicts, following treatment, have a tendency to relapse to illicit drug use regardless of the form of care which has been provided, its duration or its apparent effectiveness. This being the case, the conservative approach - and the common sense approach - is to continue treatment which is effective.

Some have rejected as nihilistic the notion that a persistent risk of recidivism is inevitable. Frankly, such rejection is nonsense, and ignores the experience of all addiction treatment specialists, regardless of their individual techniques, and regardless of their locale. It is worth considering the field of alcoholism, which in many respects is analogous. Alcoholics Anonymous, the most respected voice in the field of alcoholism treatment, has as the cornerstone of its philosophy the premise that no alcoholic is ever cured, and that the illness of alcoholism persists even after a decade or more of total abstinence. The universal experience with narcotic addiction suggests most strongly that precisely the same premise applies to that form of substance abuse as well.

And finally, a word about dosages. One of the greatest absurdities in the field of addiction treatment is the perception that there is inherent goodness in prescribing the lowest possible dose of methadone. In fact, the only consideration that matters is the effectiveness of the medication; if the patient is comfortable at a particular dosage, reports no drug craving and clinically is doing well, it is inconsequential what dosage of medication is contributing to this outcome. Thus, the answer to the question of "optimal" dosage is once again an empirical one: whatever dosage is effective! Here, too, the conservative approach is to give more rather than less, as long as no side effects occur, and indeed there are no significant side effects which have been associated with methadone in either low or high dosages.

But there is a pharmacological rationale for relying, in general, on relatively higher amounts of medication. As demonstrated in Fig. 7, the tolerance level does not rise in parallel with the methadone dosage which is administered. The individual who is maintained at 30 or 40 mgs. of methadone has little difficulty through supplemental narcotics (methadone, heroin or any other drug in this class) in reaching and exceeding the tolerance level and thereby experiencing euphoria. As the maintenance dose increases, however, the proportionate increase in the tolerance level is much greater, so that it takes steadily more and more supplemental narcotics to achieve any effect. Ultimately, at a maintenance dose of approximately 80 to 90 mgs, the tolerance level is so exceedingly high that for practical purposes it is impossible for the patient to achieve euphoria or other central nervous system effects of narcotics, regardless of the amount of additional narcotics which might be taken. The sole exception is with respect to the analgesic effect; patients are tolerant to the analgesic action of the methadone dosage being taken, but this tolerance level can be exceeded by normal pain-killing doses of any narcotic medication. But again, with regard to euphoria, the patient is pharmacologically unable to achieve euphoria through misuse of narcotics, and ultimately will not be tempted to do so regardless of the circumstances or opportunities which might arise.

## Conclusion

It must be stressed again that none of the above pharmacological phenomena are speculative or theoretical or the focus of controversy among knowledgeable individuals. They reflect what is known to every doctor, nurse and medical student who has prescribed narcotics for pain medication or for any other purpose. To summarize: methadone maintenance is associated with no euphoria whatsoever, and indeed at appropriate dosages renders the patient pharmacologically unable to achieve euphoria even with supplemental narcotics. It is safe, effective and in tremendous demand by addicts throughout the world who are motivated to give up illicit narcotic addiction and the associated life style, even though they will also give up the euphoria which they once enjoyed.

It is inappropriate to compare different treatments for narcotic addiction and to try to determine which is "better" than another. The problem is complex and difficult, and any treatment which offers help and which is acceptable to addicts should be made available. There is no question, however, that only methadone treatment lends itself to implementation on a massive scale. Any country which refuses to include methadone treatment as part of the therapeutic armamentarium is guaranteeing that only a tiny percentage of those who need and who want treatment will get it. Rejecting methadone treatment, therefore, amounts to dooming the vast majority of addicts, including those desperately motivated to give up the habit, to continued illicit drug use causing incalculable detriment to themselves and to the general society.

# 14

# Reducing Heroin Consumption During Methadone Treatment and Limitation of Post Treatment Relapses: Two Crucial Public Health Problems

**J. J. Déglon**
The Phenix Foundation, Geneva, Suisse

**Summary**

In most of Europe, drug addiction has become one of the basic public health issues with a steep rise in delinquency, prostitution, social decline, the sequels of overdoses and the transmission of Aids.
Methadone treatment programs accompanied by a therapeutic contract and psycho-social support constitutes the most effective approach for limiting the risks of addiction, decreasing the frequency of heroin use and maintaining addicts at a good psycho-social level of stability.
The results of an evaluation concerning 280 drug addicts in three methadone treatment programs of the Phenix Foundation in Geneva are presented: characteristics of the patients, decrease of heroin use, limitation of the number of arrests and imprisonments, need for an adequate dosage of methadone adapted to the needs of each patient, usefulness of maintaining the methadone dosage at a relatively high level, etc. The problem of weaning from methadone is a crucial matter for therapists. Various statistics indicate that a great majority of patients relapse within the year following the withdrawal from methadone. To limit these relapses it is necessary to wait until patients are perfectly stabilized from the medical, psychological, social, affective and toxicological point of view before encouraging them to wean themselves with a dose reduction of about 2,5 % every two weeks and with the possibility to re-adjust the methadone dosage if the need arises.

In most of Europe, drug addiction has become one of the basic public health issues and a major concern for the public authorities.

Because of a shortage of openings suiting their potential and needs, the vast majority of greatly dependent drug addicts are unfortunately condemned to vegetate in the jungle of drug addiction, with its inherent risks of medical and psychic degradation. From an epidemiological point of view this constitutes a dangerous situation, for two main reasons:

1) public health is increasingly threatened by a steep rise in delinquency, prostitution, social decline, the sequels of overdoses and the transmission of AIDS among these patients.
2) most of these drug addicts are compelled to sell heroin to yet others. In recent years it has become cheaper while its quality has improved; a sure sign that supply exceeds demand. These transactions born out of despair evidently create new addicts giving rise to an extremely vicious circle.

What solutions are available to solve this acute public health problem? There are few and there is, alas, no miracle solution. Although the results of the efforts made until now may look disappointing, I do not feel that the liberalization of narcotic drugs, which would entail a host of even more difficult new problems, is a solution.

We must keep in mind that these chronic drug addicts have, as time goes by, no real freedom of choice. They are prisoners both of the drug and of society, which locks them up in ghettos because adequate treatment and possibilities of social and professional reinsertion into society are lacking. In the long run, and whatever the psychological and social problems prior to the addiction may have been, heroin modifies durably the functioning of the neurobiological system regulating the endorphins of the brain, and this to such an extent that drug addicts must be considered as ailing people, prisoners of their own obsessive dependence and often also in a deep state of depression. This paralyzes their will to change and deprives them of the ability to make any other choice but to continue in their addiction which, in the process, gets increasingly severe.

If we really intend to break this vicious circle, so disastrous for public health, we must realize that each drug addict treated efficiently means one dealer less. He will no longer be a carrier of contamination, of infectious diseases like AIDS, and no longer proselytize or pass on narcotic drugs.

We must therefore make available to drug addicts an adequate number of treatment facilities which should not only be diversified, but also correspond to their needs and their real possibilities. These treatments should also be able to stabilize them and, above all, make them persist with the treatment until such time as when they can, eventually, stay abstinent on their own.

In-house institutions and therapeutic communities have proved their usefulness and should be supported. The evaluations made have, alas, revealed that they attract and account for only 5% of resident addicts. Moreover, most of the latter relapse once they have regained their natural environment.

A larger number of drug addicts, not exceeding 20%, can obtain conventional out-patient or in-house withdrawal treatment. These short-term cures lasting one to two weeks, have unfortunately revealed themselves as inadequate. The endorphinic systems of the brain, upset by the chronic intake of heroin, will take months to get stabilized, a fact that explains the gravity of the difficulties, known as the deficit syndrome, that arise after physical weaning. The acute anxiety, the sad moods, asthenia, psychological difficulties, sleeping problems, the inability to concentrate, are too difficult to bear in comparison with the wonderful equilibrium dispensed by heroin.

The long lasting persistence of powerful conditioned reflexes toward heroin also explains the practically inevitable relapse of addicts after any brief weaning cure. This is true to such a point that, I now believe, it is dangerous to simply propose a short-term treatment to a seriously dependent heroin addict if this treatment is not followed by a specific, lengthy post-cure period. It is like throwing a punctured life-belt to someone who is about to drown. Disappointed by a false promise of succor, that person will be even more inclined to give up.

Methadone treatment programs based on real compliance accompanied by a therapeutic contract and psycho-social support have been evaluated for the last 30 years and have shown that this constitutes the most effective approach for limiting the risks of addiction, decreasing the frequency of heroin use and maintaining addicts at a good psycho-social level of stability. Over 3000 scientific studies and evaluations have been published up to now regarding such methadone treatments. They have irrefutably demonstrated their value for treating greatly addicted, dependent heroin addicts. These treatments, alas, still meet with much opposition in some countries. This hampers their expansion, for instance in France, where at present only one addict out of 10,000 has access to a treatment facility, whereas in Geneva the proportion is 1 out of 2.

Let us discuss for five minutes the whys and wherefore of this controversy. The first misunderstanding is the customary belief that methadone lowers alertness, slows down psychomotor reflexes, diminishes intellectual capacities or the faculty of self-expression. Dr. Claude Olivenstein even states in his book "There are no happy addicts", I quote: "To distribute methadone to drug addicts is to simply transform them into medically legal invalids". Such a claim, i.e. that methadone lobotomizes drug addicts, nowadays makes every clinician who is in contact with his patients laugh. But this statement has unfortunately buttressed a prejudice shared by thousands of people. The pharmacological reality is that methadone, when administered at the correct dosage to opiate-dependent subjects, loses all its narcotic, euphorizing, sedative and analgetic effects. Contrary to what everybody is inclined to believe, drug addicts do not 'get a kick' out of methadone. It simply enables them to stay normal.

The second misunderstanding is due to the fear of unfair competition through methadone, which supposedly attracts addicts to the detriment of more demanding ways of treatment. This fear partly explains the hostility towards methadone on the part of those in charge of institutions. They fear that their staff will dwindle, state subsidies will decrease, their centers will be closed and that unemployment will ensue. Quite the contrary has happened: the more methadone treatment centers there are, the more requests for admission are recorded by traditional treatment centers.

A third misunderstanding is caused by the mistaken belief that methadone is a life-long replacement treatment. We shall subsequently see that a certain number of patients are capable, after several years of methadone treatment, of undertaking on their own a very gradual withdrawal without relapsing into their addiction.

The second reason for the opposition of many specialists to methadone treatment is due to the poor results obtained with certain programs. Can one condemn a surgeon if he is being sent inoperable patients only, if he is compelled to administer only half a dosage of anesthetic, or if the time he can devote to an operation is severely curtailed? The indications for methadone treatment must be properly determined, the dosage be continuously adapted to the needs of each patient and the cure maintained for the time required. To allocate to our Parisian colleagues but 50 of the most debilitated cases from among the thousands of drug addicts in that city, is nothing less than a hyper-negative selection, which in no way allows the efficacy of these treatments to be judged. To diminish the methadone dosage in order to punish a patient each time he has used heroine again or to maintain him artificially at a sub-dosage level - a phenomenon all too frequently encountered in certain programs - is a pharmacotherapeutic error equal to giving a diabetic a sub-dosage of insulin or to prescribing too little thyroid hormone for a hypothyroid patient. Destabilized by an excessively low methadone dosage, the drug addict will of course try to alleviate his physical suffering and his psychological discomfort by using heroin, which will very soon seriously interfere with his treatment. Witnessing this vicious circle does not justify the conclusion that methadone cures are inefficient, for they require great medical know-how to produce results that will be the better the more they are accompanied by specialized psycho-social support. For an improved comprehension of all these misunderstandings, it is important to stop "demonizing" methadone and to release it from its limbo. Thousands of addicts, whom we must also consider as human beings and who at present have no access to existing therapeutic treatments, would thus be able to benefit at long last from treatments adapted to their specific needs.

The most seriously perturbed drug addicts unfortunately cannot respect the minimum enrolment criteria of these programs or are not attracted to them, either because methadone does not procure any pleasure or because of the bothersome bureaucratic procedures associated with these cures.

This is why, within a policy framework aimed at reducing the risks, at attracting seriously affected drug addicts to a therapeutic system and at keeping them in it, we shall experiment in Switzerland a possibility of helping them survive in a more humane way by administering, under medical control, injectable methadone, heroin and morphine to a few groups composed of some 50 individuals.

Pending the results of these much questioned experiments, I should like to analyze with you several results which will enable us to know what we can expect of methadone treatment, in particular as regards the reduction of heroin use and needle-sharing, these being the essential objectives of these treatments and a foremost imperative for public health.

This is why we shall examine some of the results obtained by the Phenix Foundation. The Phenix Foundation, which I run, is a non-profit institution, treating with methadone 280 drug addicts split up into three medico-psycho-social programs. The therapeutic team consists of 22 persons, comprising doctors, psychologists, nurses, social workers, medical assistants, etc.

In April last year, we evaluated all our programs as a whole, by interviewing our patients in depth, in order to better specify the characteristics of the individuals being treated and the results of the treatments.

Men accounted for 79% and women for 21% of the group members. Their mean age was 30 (19 to 44 years).

The average age on taking heroin for the first time was 18.9 (13 to 36 years) and the average time lapsing between the first intake of heroin and daily use was 2.4 years. We observed an average of 4.1 years of regular heroin addiction before starting methadone treatment. The average methadone dosage was 57 mg. The average number of methadone distributions per patient was 2.9 weekly. The average duration of treatment at the time of the evaluation was 3.4 years.

The number of arrests and imprisonments dropped drastically as of the start of treatment, which goes to show that drug addicts are delinquents out of necessity and not by nature. To be more precise, our patients had, before the cure began, been in prison for an average duration of 5.4 months, that is 172 days. If, in order to make a valid comparison, one takes into consideration the 4.21 years of addiction, this gives 42 days of imprisonment per year of addiction. During the year prior to our evaluation, our patients had spent an average of 3.4 days in prison, which corresponds to a 92% decrease in incarceration time. 56.5% of the patients were seropositive for B hepatitis, 24.6% had been vaccinated, 11.7% were negative and 7.3% had not been tested. 67.7% of the patients were seropositive for C hepatitis, 23.4% were negative and 8.9% had not been tested. 27.9% of our patients were seropositive with regard to the AIDS virus, 68.8% were negative and 3.2% had not been tested. Three quarters of our patients were well stabilized with methadone, 14.5% were taking anti-depressants, 5.6% tranquilizers, 2.4% sleeping pills, 2% neuroleptics and 1.6% disulfiram (Antabus) for reasons of alcoholism.

It is of interest to note that the average monthly income of our patients amounts to 3029.00 Swiss francs. 57% of these incomes derive from a declared income, 12% from an insurance, 5% from welfare, 3% from parents and 2% from friends, while only 1% was admittedly obtained illegally.

On the average, our patients have each repaid 5021 Swiss francs to the state during the last 12 months. This covers various taxes, but also debts and arrears for court costs.

This sum corresponds to the annual cost of methadone treatment (medical - including urine tests - and psychotherapeutic). In terms of the cost/yield ratio for the public, one notices that these treatments are of particular interest.

The most useful aspect, from a public health point of view, is to decrease the use of heroin to the utmost. How did our programs perform in this respect?

We have observed that during the last month before starting treatment with us, our patients used an average of 3.6 doses of heroin daily, that is over 100 a month.

Thanks to the interviews and the urine analyses carried out twice a week, we are in a position to quantify quite precisely the drugs used by our patients. By adding tracers to the methadone, which we then measure in the urine, and through analyses of the saliva, we know exactly how often they cheated with the urine, i.e. 4.6% of the samples examined during the month of evaluation. Taking into account this small margin of error, we find 1.6 intakes of heroin monthly. Of these 75% of our patients inject it, 9% smoke it and 16% sniff it. Compared to one hundred intakes of heroin taken prior to the cure, we obtain a 98% decrease in heroin consumption with our methadone patients. This also explains why these treatments are so interesting with a view to limiting the risk of AIDS. 95% of the addicts who were lucky enough to join the methadone treatment program prior to the AIDS epidemic remained seronegative. By comparison we counted over 50% of seropositive subjects among our patients during the years '84 and '85.

As regards cocaine, we observed an average of 0.2 intakes during the month preceding our evaluation. 38% of the patients injected it and 62% sniffed it.

What are the factors that further the decrease in heroin use through methadone treatment?

Firstly, it is of capital importance to maintain a methadone dosage constantly adapted to the needs of each patient. In the course of an evaluation of American programs applying to 407 methadone patients, J.C. Ball has demonstrated that the frequency of heroin use during the last month was inversely proportional to the prescribed methadone dosage. 50% of the patients whose dosages were below 20mg took heroin at least once a month, while all the patients at more than 75mg abstained completely from taking opiates.

This study demonstrates the usefulness of maintaining the methadone dosage at a relatively high level, above all with those patients who continue to use heroin.

Our evaluations also demonstrated that the most abstinent patients are those that have been under treatment for the greatest length of time, this is why it is so important to keep them under treatment as long as possible.

There is no pharmacological activity of methadone that is effective with regard to reducing the number of heroin intakes among the patients under treatment. Our experience and evaluations abroad have demonstrated perfectly the primary role of the limits and of a protective framework represented by the therapeutic contract, monitored by means of regular urine analyses. In addition to these reassuring exigencies, a warm psychological support given in the course of regular discussions and social work are also essential factors towards improving the results, which is the main objective if one wants to improve the equilibrium of our patients and public health.

**Limiting Relapses after Methadone Treatment**
**Another Imperative for Public Health**

The problem of weaning from methadone is a crucial matter for therapists. When, how and with whom should weaning be undertaken?

Various American statistics indicate that a great majority of patients relapse within the year following their weaning from methadone.

In a few weeks, these relapses can annihilate years of therapeutic efforts, seriously threatening the patients' psychological and social equilibrium as well as their health because of the threat of AIDS and overdoses. Reverting to heroin after years of efforts is like a dam that bursts: a compulsive tidal wave overwhelms the personality. Neither intelligence nor will-power are sufficient to stem the flood. A deep depression often accompanies this relapse: the subjects are profoundly disappointed in themselves and an almost suicidal behavior can be observed.

I remember a young, intelligent woman, well stabilized during her treatment, very much aware of the risk of AIDS because we had often spoken about it. She wanted to wean herself from methadone in a special medical center. A few weeks later she came to ask me to take her back into my methadone program because of a serious relapse. She then confessed to me that she had not only had sexual relations with a seropositive addict without using a condom but that she had, moreover, used his needle.

On that day I realized the limitations of education and of AIDS prevention. It is not enough to inform drug addicts seriously about the dangers of AIDS and the ways to protect themselves. Above all one must see to it that they maintain the best possible psychological equilibrium enabling them to keep their impulses in check.

That was two or three years ago; during that same period we had witnessed several tragic relapses among patients of ours who had tried to wean themselves from methadone. One blow came after another: we heard of the suicide, the overdose, the incarceration or the HIV seroconversion of one or another of these patients. I then realized, jointly with my colleagues, to what extent our weaning policy had been unsuitable and dangerous. We ourselves, maybe still feeling guilty because of long-term administration of methadone, had encouraged our patients to wean themselves off methadone as soon and as rapidly as possible. In turn, they had been strongly urged by their environment and by their own guilt-feelings to do so. It is true that drug addicts are still far much inclined to consider methadone similar to heroin. This is why the faster they lower their methadone dosage the better they feel, and feel bad when they have to increase it again. This moral value attributed to the methadone dosage is unique. For instance, a diabetic who has to increase his insulin dosage does not at all feel guilty about it. A major effort must be undertaken towards informing patients and their families. We must move methadone out of the drug addiction and moral standards area and relocate it in its medical and neurobiological ambit.

We have now understood that the methadone dosage should be constantly adapted to the specific needs of each patient. I wish to remind you that methadone is a very well-tolerated substance and among those studied in greatest depth in the history of medicine. Over 3000 scientific works and assessments have not been able to demonstrate seriously any evidence of the toxicity of methadone for the body, even when taken continuously and at high dosages.

The only criterion that we must opt for is therefore that of the quality of life of our patients, which we must maintain at all costs, with or without methadone.

Is it possible, though, to wean drug addicts off methadone while enabling them to maintain the quality of life acquired during the cure? If yes, with which patients, and when and how? Several years ago, thinking of one of my patients, a young bank employee who wished to wean himself, I put this question to Robert NEWMAN, one of the top American experts. NEWMAN asked me: "Is he well stabilized?"

"Yes, I said."

"Has he abstained from opiates for more than one year?"

"Yes."

"Does he maintain a stable affectionate relationship with a girl friend?"

"Yes."

"Has he paid his debts?"

"Yes."

"Does he have a satisfactory job?"

"Yes."

"Has he resumed normal leisure activities?"

"Yes."

At that NEWMAN looked and me and asked: "Well, if he is doing fine, why do you want to bother him?"

With this provocative answer, all my colleague wanted to tell me was: "Be content with your success and, above all, do not touch the methadone dosage unless you want to upset everything."

Well, does that mean that we should keep drug addicts on methadone throughout their whole lives? We ourselves are not quite convinced of this, for we have fortunately been able to wean successfully dozens of patients who appear not to have relapsed for several years now.

We envisage an extensive scientific evaluation in the next two years, in order to find out exactly how our former patients are doing.

It must be said that NEWMAN was heavily influenced by the study he had carried out in HONG KONG in 1972 and which has remained a pilot study.

His objective at the time was to measure the pharmacological action of methadone in respect of the therapeutic factors of the program, such as the framework, urine checks, psychotherapy, social assistance, etc. Opponents had said to him: "After all it's not the methadone that is useful, it is the psycho-social treatment. Give them orange juice instead of methadone and they will do just as well".

To check this hypothesis, NEWMAN arranged with the pharmacist who prepared the methadone dosages for a program involving several hundred drug addicts in Hong Kong to wean half of them by 1 mg daily, using a double blind control, without the therapists' knowledge.

The fifty subjects of the control group could freely modify their dosages, which were between 30 and 120 mg with an average of 97 mg. The patients of the second group received a methadone does that was reduced by 1 mg per day even when they asked to have their dosage increased.

For both groups, the criterion for exclusion from the study was dropping out of the treatment by the patients and the occurrence of six positive opiate urine checks running.

The results obtained were impressive. In the group being weaned, after 32 weeks and in spite of the efforts made by the therapists to stabilize them, only 5 patients remained out of the initial 50, that is 10%. Two thirds had been excluded because of heroin relapses and one third because they dropped out. In the control group, where dosage was free, 76 were still on the program after 32 weeks. After exclusion from the program, patients were re-admitted to the methadone program at a normal dosage and their retention rate remained the same as that of the control group.

In 1977, SENAY has confirmed the importance of a very slow reduction of methadone in order to avoid relapses. He chose 127 volunteers who had been in the treatment program for over one year, whose methadone dosage had been stable for at least one month, and who had not had any positive urine tests during the last month of treatment and no more than two positive tests in the course of the last six months. He divided these 127 patients up into three groups according to the double blind technique. Neither the patients nor the therapists knew the dosages. The group A patients (control group) stayed at a normal dosage. The group B patients were being weaned rapidly in steps of 10% of the weekly dosage. The group C patients were weaned slowly, by about 3% of the weekly dosage. After 30 weeks 94% of control group A were still on the program, but only 24% of group B and 53% of group C which had been weaned slowly. One can therefore observe the following fact: in spite of slow weaning by about 3% per week, half of the patients relapsed within 8 months.

Let's take a look at some interesting findings resulting from our three treatment programs at the Phenix Foundation.

During the last few years we have admitted between 80 to 100 drug addicts annually, specifically 109 and 110 respectively in 1991 and 1992. In these groups one must take into account the new patients who had never undergone any treatment at the Phenix Foundation and the earlier patients who had weaned themselves off methadone before but had asked to be re-admitted because of relapses. It is interesting to note that in these last five years, there has been a spectacular and regular decrease of the number of relapses among our former patients.

In 1988, 53%, that is to say over half of the admissions referred to former patients, while in 1992 there were only 10% of relapses.

How can this finding be explained? Our former patients know that in case of need they may resume treatment with us at any time, short-cutting the waiting list, to avoid losing all their previously acquired benefits. We therefore hold that pending a scientific evaluation, our weaning indications are better specified, and that we regularly improve the quality of these weanings. This explains the regular reduction of relapses observed.

What is this new policy concerning weaning? To begin with, we wait until patients are perfectly stabilized from the medical, psychological, social, affective and toxicological points of view before encouraging them to wean themselves, if they so wish and feel fully motivated to do so.

We suggest to the patient a dosage reduction of about 2.5 % every two weeks, with the possibility for him to re-adjust his methadone dosage at any moment for the length of time required, if the need arises.

Family, marital or professional problems are likely to interfere seriously with the treatment. In the event that the patient's quality of life is being upset, which may emerge in the form of anxieties, sleeping problems, a state of irritability, a tendency to get depressed, an increase in alcohol consumption, use of heroin, cocaine or tranquilizers, we then re-equilibrate the methadone dosage immediately to prevent any relapse.

With regard to some patients, who, after two or three attempts at weaning, that were suspended because their equilibrium had become upset, we hold that methadone represents a psychotropic medication that is necessary to maintain durably their quality of life. These patients must be considered as "methadonics", on a par with diabetics, manic depressives or epileptics, who must take medication regularly in order to stabilize their chronic disease.

It matters to accept these methadonics, who have nothing in common with drug addicts, as fully normal individuals, and a stop must be put to the all too frequent discrimination against them.

To conclude, it is in the interests of the drug addicts themselves and of public health, that, in order to avoid the disastrous effects of weaning from methadone, the indications be very carefully determined and weaning proceed very slowly and digressively, giving the patients every latitude to re-establish their equilibrium for the period of time required, should they feel the need to do so.

## References


1. Ball J C. Reducing the risk of AIDS through methadone maintenance treatment. Journal of Health and Social Behaviour 29:214-226, 1988.
2. Ball J D, Ross A. The effectiveness of methadone maintenance treatment, Springer-Verlag, New York, 1991.
3. Batki S. Treatment of intravenous drug users with AIDS: the role of methadone maintenance. J Psychoactive Drugs 20:213-216, 1988.
4. Bayer R, Koenigsberg L. Retention rates among methadone patients: an analysis of the New York experience. 1964-1976 International Journal of 16:33-41, 1981.
5. Briefer J F, Déglon J J. Traitement de méthadone et comportements à risque de transmission

HIV. Méd et Hyg 50:1544-1549, 1992.
6. Caplehorn J R M, Bell J. Methadone dosage and retention of patients in maintenance treatment. The Medical Journal of Australia 154:195-199, 1991.
7. Cooper J R. Methadone treatment and acquired immuno-deficiency syndrome. J Am Med Assoc 262:1664-1668, 1989.
8. Dole V P. Addictive behaviour. Scientific American 243:138-154, 1980.
9. Dole V P. Implications of methadone maintenance for theories of narcotic addiction. J Am Med Assoc 260:3025-3029, 1988.
10. Dole V P, Joseph H. Long term outcome of patients treated with methadone maintenance. Ann N Y Acad Sci 311:181-189, 1978.
11. Dole V P, Kreek M J. Methadone plasma level: Substained by a reservoir of drug in tissue. Proc Natl Acad Sci USA 70:10, 1973.
12. Dole V P, Nyswander M E. Methadone maintenance and its implication for theories of narcotic addiction. In: The Addictive State, edited by Wikler A. Williams and Wilkins, Baltimore, 1968, pp 359-366.
13. Dole V P, Nyswander M E, Kreek M J. Narcotic blockade. Arch Intern Med 118:304-309, 1966.
14. Déglon J J. Le traitement à long terme des héroinomanes par la méthadone, Editions Médecine et Hygiène, Geneve, 1982.
15. Hunt G H, Odoroff M E. Follow up study of narcotic drug addicts after hospitalization. Public Health Rep 77:41-54, 1962.
16. Inturrisi C E, Vereley K. The levels of methadone in the plasma in methadone maintenance. Clin Pharmacol Ther 13:633-637, 1972.
17. Joe G W, Simpson D D. Retention in treatment of drug abusers: 1971-1972 DARP admission. Am J Drug Alcohol Abuse 2:63-71, 1975.
18. Kreek M J. Medical safety and side effects of methadone in tolerant individuals. J Am Med Assoc 223:665-668, 1973.
19. Kreek M J. Plasma and urine levels of methadone. N Y State J Med 73:2773-2777, 1973.
20. Kreek M J. Medical complications in methadone patients. Ann N Y Acad Sci 311:110-134, 1978.
21. Kreek M J. Rationale for maintenance pharmacotherapy of opiate dependence. In: Addictive States, edited by O'Brien C P, Jaffe J H. Raven Press, New York, 1992, pp 205-230.
22. Mino A. Analyse scientifique de la littérature sur la remise controlée d'héroine ou de morphine, OFSP, Berne, 1990.
23. Newman R G. Narcotic addiction and methadone treatment in Hong Kong: Lessons for the United States. J Public Health Policy 6:526-538, 1985.
24. Novick D M, Joseph H, Croxson T S. Absence of antibody to human immunodeficiency virus in long-term, socially rehabilitated methadone maintenance patients. Arch Intern Med 150:97-99, 1990.
25. Simpson D D, Savage L V, Lloyd M R. Follow up evaluation of drug abuse during 1969 to 1972. Archives of General Psychiatry 36:772-780, 1979.
26. Stimmel B, Goldberg J, Cohen M, Rotkopf E. Detoxication from methadone maintenance: Risk factors associated with relapse to narcotic use. Ann N Y Acad Sci 311:173-180, 1978.
27. Zweben J E, Payte J T. Methadone maintenance in the treatment of opioid dependence: A current perspective. The West Journ of Med 152(T):588-599, 1990.

# 15

# Methadone Against AIDS. The Swedish Experience

**O. Blix**

Department of Alcohol and Drug Diseases, Malmö General Hospital, Sweden

## Summary

After a period of 15 years as a local experiment in Uppsala, followed by 5 years of closed admission to the program, Methadone Assisted Rehabilitation is now an accepted, regular treatment for heavy heroin addiction in Sweden. More than 600 patients have been or belong to the 423 who are still in treatment . Some have successfully left treatment. Some 150 have been involuntary detoxified and most of those have relapsed into heroin addiction; some are back in treatment again. Others are still abusing, a few have become drug free and some have died.

Over all, when using comparatively strict inclusion criteria and when keeping strict criteria for staying in treatment, the MAR model gives a dramatical effect on survival and rehabilitation compared to drug free treatment or no treatment. Results from other models of methadone treatment, such as low threshold or harm reduction programmes do rarely produce such clear results. This can be explained by many factors. One is a tendency to under medicate (prescribe an insufficient dose) in these programmes. Others have to do with patients expectancy and motivation and treatment resources. If the program goal is to reduce the harm of addiction to the individual, but not to rehabilitate back to society, one should not expect any other results.

In Sweden, we have chosen a high ambition for our treatment, which gives high outcome measures. But we may loose patients on the way, who might have survived in a harm reduction program in Holland or Liverpool.

## Introduction

The medical use of opiates for pain relief has served humanity since thousands of years. The addictive potential has also been known, and has caused individual tragedies as well as serious societal damage.

The drug free treatment of chronic opioid dependence is difficult and has generally shown a low efficiency and high cost. When Drs Dole and Nyswander in the nineteensixties developed the modern, long term substitution therapy with methadone and psychosocial rehabilitation in a comprehensive setting, a new era in the treatment of drug addiction was created. The method has been developed in many countries, in many programmes and the literature is extensive today, proving the efficacy and safety of comprehensive methadone maintenance treatment.

## Sweden

Sweden has about 8,5 million inhabitants. The drug problem has developed from a rather minor one until the 60´ies to a more serious situation today. Amphetamine is the most common street drug of injection. Heroin was introduced on the illicit market around 1974. Since the start of Methadone Maintenance Treatment (MMT) in 1966 the drug scene has changed.

In 1968, a case-finding survey found 200 individuals abusing prescribed opiates or raw opium and morphine base. In these days, drugs were cheap and thus few were involved in criminal activity. In 1979, the number of opiate addicts had risen to 3-4000. This number has been fairly constant until recently. The same study estimated the total number of heavy (= daily and/or i.v.) drug abusers to be 11.000-14.000.

However, reports today from the police, customs, social workers and detoxification units indicates a new wave of smoke heroin abuse among persons in the age of 15-25 years. The estimated total number of addicts is unchanged, but there is a difference from 1979 in that illicit drugs can be found in most areas of the country today. Earlier it was a typical city problem. Heroin is still mainly abused in the greater Stockholm area and in the south of Sweden.

Most heroin addicts entering Methadone programmes today have experienced long term imprisonments, several intoxications and infections and they are rarely employed or educated. Several are HIV positive. Concomitant psychiatric disease is common as well as poly drug abuse.

## Swedish Methadone Treatment History

When the first study on MMT was published by Dole and Nyswander (1965), Prof. Lars M Gunne from the Uppsala University, was a guest investigator at Dr. Doles laboratory at the Rockefeller University in New York. He brought the new method back home and started the first Swedish methadone program, later developed into Methadone Assisted Rehabilitation (MAR).

The rationale for this treatment is the effect of long term stabilization on methadone, which enables the drug addict to concentrate on social, physical and psychological rehabilitation. It also enables the therapist to diagnose and treat other disorders or diseases.

My own participation as a doctor in the Swedish MAR started in 1978 at Prof. Gunnes clinic in Uppsala, and I will today summarize some of my experiences and conclusions from the years I have had the privilege to work with this treatment, in Uppsala, Stockholm and Malmö.

## MAR

The development of the MAR in Sweden is a result of the extensive research by prof. Gunne and his coworkers at the University of Uppsala. Their experiences have laid the foundations for the three new programmes, that have opened during the last decade in Stockholm (1986), Lund (1990) and Malmö (1992). The new programmes have tried to keep a similar policy towards the patients, but naturally, differences have developed. To avoid too big differences, regular meetings with participation from all programmes and other representatives from the treatment of drug addicts are hosted by the Board of Health and Welfare (BHW) in Stockholm. There is also an ongoing exchange of experiences between the programmes through mutual visits and meetings. Recently, a large scale evaluation was proposed from the BHW, which is partly inspired by the evaluation by Ball and Ross (1991).

### *Inclusion Criteria*

The criteria for entering MAR treatment have been unchanged since the start in the sixties until 1989, when the need of 3 earlier treatment attempts was abolished. Today the criteria are:

1. Minimum age 20 years.
2. Current i.v. opiate abuse.
3. A documented history of at least 4 years of compulsive opiate addiction.
4. Voluntary application. No pending charges.
5. No dominating side abuse.

Earlier treatment experiences are recommended before applying to MAR .

Exclusion Criteria

-Drug trafficking and/or diversion of Methadone.
-Violence or serious threat of violence.
-Criminality committed during MAR treatment leading to imprisonment.
-Patients are also excluded in case of prolonged side abuse or continuing compulsive opiate abuse.

## HIV

HIV started to spread in the i.v. drug user population in Sweden in 1983 according to analysis of frozen sera at the national bacteriological laboratory in Stockholm. At that time the MAR program was still closed for admission due to political reasons. In a study by myself and Grönbladh, presented at the first European Symposium on Drug addiction and AIDS in 1991, it was found that the HIV prevalence among patients who had entered the Swedish MAR program before 1983 was 3 %, whereas the prevalence increased among addicts entering the program the following years. It peaked at 59% in 1988. Furthermore, those patients who were HIV negative upon admission stayed negative throughout the studied period (1984-1991). Today, the prevalence is around 15% among new admissions (Fig. 1).

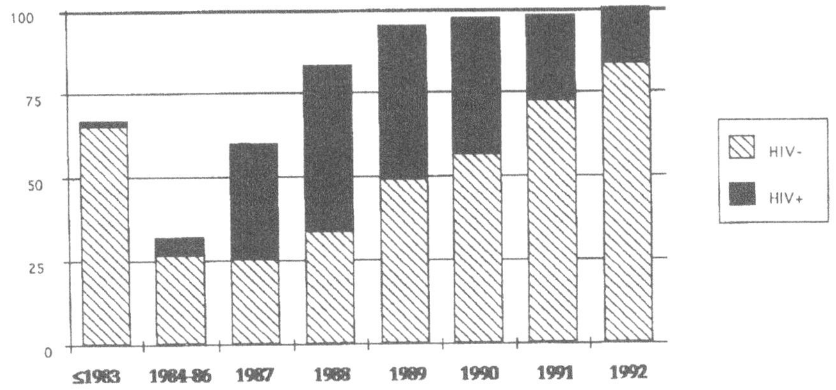


**Fig.1.** Total number of admission to MAR in Sweden and number of HIV positive admissions per year

The total number of HIV positive patients receiving methadone treatment today is 112, of whom 100 are treated in Stockholm, 2 in Lund and 10 in Uppsala. The total number of known HIV infected drug addicts in Sweden was 648 in June 1993.

## Uppsala

Some of the unique features of the original Uppsala Program has to do with its status of being a program with patients scattered around a large area, some of them living 700 or more km away from the clinic. Thus the program have had to develop close cooperation with local treatment facilities such as general practitioners, nurses, social workers and last, but not least, the local pharmacies. The latter are dispensing the daily methadone dose to patients as soon as they are discharged from in patient status. They also send the urine specimen to the laboratory in Uppsala for analysis.

The selection of patients into MAR was in Sweden done by the Uppsala Program only until 1988. At that time the Stockholm program took over this service for their own catchment area. Today the Uppsala program serve the rest of Sweden, including the two programmes in the south in Lund and Malmö.

The selection is done out patient by an evaluation team. The initiation of methadone medication is still in most cases done in-patient. After an average of 3 months, the patients return to their homes, and get their mediation dispensed through a local pharmacy. The out patient treatment is handled by a team of two social workers, one physician, one half time psychologist and a supervisor (L Grönbladh), who is also collecting data for evaluation. Professor emeritus Lars M Gunne is responsible for the research, and is presenting recent data at this symposium.

Several important studies have been presented throughout the years, proving the efficacy of the MAR in saving lives, reducing criminality, reducing prostitution, reducing the spread of AIDS and prolonging life expectancy among treated patients. As being one of the oldest methadone programmes in Europe, they have collected a unique knowledge and experience of this treatment during their 27 years in operation.

**Stockholm**

In 1985, the county of Stockholm decided to open an out patient clinic for methadone patients living in Stockholm, but received treatment from Uppsala. The clinic was designed to take care of the 40-50 patients that were already in treatment. However, when the clinic opened in January 1986, the HIV epidemy among heroin addicts in Stockholm resulted in an increasing number of referrals to MAR, and soon the Stockholm program had to increase its capacity. The initiation of MAR was then started in out patient status, which is a model widely used in other countries since many years. It proved to be almost as safe as the in-patient initiation, though problems arose when patients started to abuse other drugs and needed to be detoxified from this side-abuse. Thus, in 1990, a ward was opened as well in Stockholm.

There are separate out patients units serving about 50 patients each. Two of these units are specialized on HIV positive and AIDS patients. Two therapeutic communities have also been started through the Stockholm program. These treatment facilities are today more independent and accept patients from all MAR programmes in Sweden.

A group consisting of the head of the program (Stefan Borg, MD), a special investigator, social workers from the city of Stockholm, the responsible MD:s from the three detox clinics in Stockholm and representatives from drug free treatment are evaluating the patients applying for treatment according to the Swedish inclusion criteria.

Today, 273 patients are treated within the Stockholm program. Of those 100 are HIV positive. A total of 78 patients have been involuntary detoxified from methadone treatment between the start in 1986 until June 10 1993; 40 of those were HIV positive. Most of those are waiting to be, or readmitted again to the program.

In Stockholm, the controlling aspects of the program has been stressed, and a qualified laboratory for detection of even small amounts of drugs in urine has been set up. Studies have been performed on the plasma levels of methadone with the separation of active and inactive isomers on patients with craving, despite normal plasma levels of the racemic methadone. In some cases, it is shown that these patients can have a very low concentration of the active L-isomer whereas the D-isomere is high. The reason for this is not clear, but it may explain why sometimes there is a discordance between plasma methadone levels and performance in the treatment.

**Lund**

In Sweden, the majority of heroin addicts live either in the greater Stockholm area or the southwest, i.e. Malmö-Helsingborg. Lund is a university town, located between Malmö and Helsingborg, where treatment of drug addicts has a long tradition. The unit has been lead by Dr. Kerstin Tunving, who have over the years gained a reputation for her intrepid actions for drug addicts living in the south. She has initiated the needle exchange program in collaboration with the department of infectious diseases in Lund in 1987, which in a few months inspired the colleagues in Malmö to do the same. In 1990 she opened the third methadone program in Sweden. Today, this out patient facility accommodates 56 patients, and there are plans to open an in patient ward shortly. Most of the patients come from Helsingborg, a 45 minutes ride away by train.

**Malmö**

The latest program to open in Sweden was the one in Malmö, which has now been in operation since February 1992. The reason to open this program was that Malmö, the third largest city in Sweden, has a large population of heroin addicts, and the costs for treating those in Uppsala and Lund were rapidly increasing. It was estimated that for the same cost, twice as many patients could get as good or better treatment if a local program was opened. So far there is nothing that has contradicted this assumption. The program is also an out patient unit staffed with two psychiatrically trained nurses with experience from the detoxification unit in Malmö and Lund, one social worker with long experience and a secretary, who also have worked within the detoxification unit for many years. The physician was recruited from the Uppsala program.

Today some 31 patients are in treatment. Of those, 52% hold down a work or study full time. The drug situation is stable, i.e. the majority are doing well or very well (no side abuse or irregular abuse) whereas a small group (5 individuals) are more or less continuously taking other drugs than methadone. Four of those are now withdrawing from methadone medication.

Concerning the difficult vocational situation in Sweden, with an unemployment rate around 10% (in Malmö it is 13%), it is satisfying to find around 50% of the patients in work or school after only a year in treatment. The other programmes in Sweden have similar results for their clientele. It is now increasingly difficult to find solutions on the vocational problems, both because of the general situation in the country, but also because many of the patients need a long time to treat psychological and physical disorders. The cooperation with the social workers has become less efficient as their resources have been drastically cut the last year. But the earlier controversy between social workers and the medically oriented treatment of drug addicts is no longer a problem. Hopefully as a result of several successful cases of cooperation with patients and information.

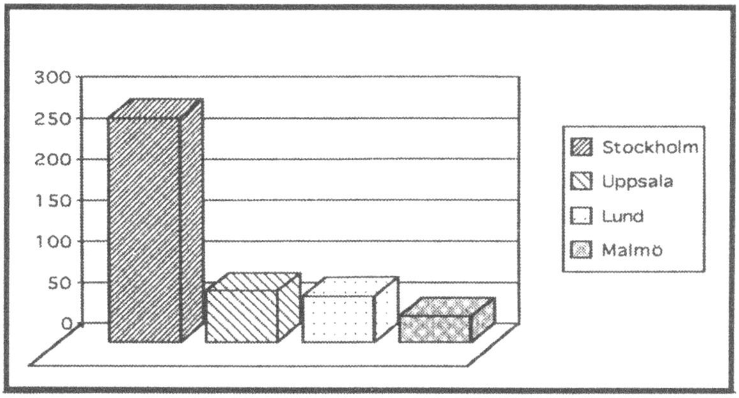


**Fig. 2 .** Total number of patients in MAR treatment in Sweden, Sept. 1, 1993

## Conclusions

A comprehensive treatment model has to provide a safe and correct dispensation of methadone to patients who are in need of this treatment. This implies:

1. The use of eligibility criteria to select patients with a documented history of heavy addiction to opiates.
2. Provision of other services than the sole dispensation of methadone once patients are in treatment.
3. The use of quality assurance. In the treatment, i.e. continuous follow up of the effects of the treatment (for example through regular drug testing of urine, home visits to patients, check up of vocational status and physical as well as psychiatric health status) and adequate dosage.
4. Integration in the society. Ability to cooperate with the patients and their families, with co-workers dealing with the same patients from other organizations like the criminal justice system, social welfare system, the educational system and employers. Network.
5. The use of criteria for staying in treatment (discharge criteria). This in turn implies the need of rules. Which have to be few, simple and easy to interpret.
6. Clear, and individually realistic goals for the treatment. This could include ultimately detoxification from methadone, but should not be mandatory.

There are also components in treatment that can't be fully verbalized, but probably are as important as any other to predict successful outcome. It has to do with "compassion", "humanity", "dignity", "empathy", "respect" and "caring". It has to do with the over all goal to build the self esteem of the patients, so that they can mature to individuals with a responsibility for their own lives and for their treatment as long as they need it.

After a period of 15 years as a local experiment in Uppsala, followed by 5 years of closed admission to the program, Methadone Assisted Rehabilitation is now an accepted, regular treatment for heavy heroin addiction in Sweden. More than 600 patients have been or belong to the 423 who are still in treatment (Fig. 2). Some have successfully left treatment. Some 150 have been involuntary detoxified and most of those have relapsed into heroin addiction; some are back in treatment again. Others are still abusing, a few have become drug free and some have died.

Over all, when using comparatively strict inclusion criteria and when keeping strict criteria for staying in treatment, the MAR model gives a dramatical effect on survival and rehabilitation compared to drug free treatment or no treatment. Results from other models of methadone treatment, such as low threshold or harm reduction programmes do rarely produce such clear results. This can be explained by many factors. One is a tendency to under medicate (prescribe an insufficient dose) in these programmes. Others have to do with patients expectancy and motivation and treatment resources. If the program goal is to reduce the harm of addiction to the individual, but not to rehabilitate back to society, one should not expect any other results.

In Sweden, we have chosen a high ambition for our treatment, which gives high outcome measures. But we may loose patients on the way, who might have survived in a harm reduction program in Holland or Liverpool.

## References

1. Blix O, Gronbladh L. The impact of methadone maintenance treatment on the spread of HIV among i.v. heroin addicts in Sweden. In: Drug Addiction and AIDS, edited by Loimer N, Schmid R., Springer Verlag, Wien, 1991.
2. Gronbladh L, Ohlund L S, Gunne L M. Mortality in heroin addiction: Impact of methadone treatment. Acta Psych Scand 82:223-227, 1990.
3. Dole V P, Nyswander M E. A medical treatment for diacetylmorphine (heroin) addiction: A clinical trial with methadone hydrocloride. J Am Med Assoc 193:80-84, 1965.
4. Ball J C, Ross A. The effectiveness of methadone maintenance treatment, Springer-Verlag, New York, 1991.
5. Bottinger M, Biberfeldt G, et al. HIV testing among injecting drug users in the Stockholm area; information compiled from testing laboratories. Scand J Inf Dis 25:289-295, 1993.
6. Gronbladh L, Gunne L M. Methadone-assisted rehabilitation of Swedish heroin addicts. Drug Alc Dep 24:31-37, 1989.
7. Gunne L M, Gronbladh L. The Swedish methadone maintenance program: A controlled study. Drug Alc Dep 7:249-256, 1981.
8. Olsson B, Carlsson G, Fant M, Johansson T, Olsson O, Roth C. Heavy drug abuse in Sweden 1979 - A national case finding study in Sweden. Drug Alc Dep 7:273-283, 1981.

# 16
# Addiction and Pregnancy: Maternal and Child Issues

**L. P. Finnegan**
Women's Health Initiative, National Institutes of Health, Bethesda, MD, U.S.A.

## Summary

The treatment of drug-dependent women presents challenges for the women as well as treatment professionals. A number of barriers exist that may prevent women from enrolling in treatment. These barriers include the paucity of gender specific services and those that are ethnically and culturally sensitive. Moreover, drug dependent women tend to neglect general health care and management is difficult due to frequent medical and psychiatric complications including the risk for HIV/AIDS. Socioeconomic difficulties such as poor housing, inadequate income, and lack of education, are complicated by emotional problems such as feelings of worthlessness and depression. Treatment for drug dependent women must include services for their medical illnesses, psychosocial counseling, and addictive therapies utilizing various modalities such as methadone maintenance and drug free in-patient services. Assessment of parenting skills in women with dependent children allows the identification of mothers who may be at risk for poor parenting, child abuse, or neglect. The potential complications associated with substance abuse can be markedly diminished in women if treatment professionals show compassion and provide persistent, comprehensive services.

Throughout the world, millions use illicit drugs regularly. Many millions more are addicted to nicotine, alcohol, or both. Large numbers of people die every day as a result of nicotine's role in heart disease, lung disease, and cancer. The effects of alcoholism have wreaked incalculable damage across generations in our society.

The epidemic of drug abuse has overwhelmed women and children. The many negative effects of drug exposure on newborns are familiar to perinatologists and neonatologists (1-8). The AIDS epidemic complicates this picture. Newborns who test positive for human immunodeficiency virus (HIV) usually have parents who were drug abusers or whose partners were drug abusers (9). Many of these infants are "boarder babies" living in large city hospitals, physically impaired, and abandoned by their parents. These babies will be placed in foster care, when it is available, away from their families, immediately on leaving the hospital (10).

The tragic impact of alcohol on infants and children is well known. The incidence of fetal alcohol syndrome is estimated to be one to three cases per 1,000 live births, or about 4,000 to 12,000 new cases in the United States each year. This preventable disease is one of the leading known causes of mental retardation (11).

Within the context of comprehensive services for pregnant drug-dependent women over the last 20 years, we have learned a great deal through clinical observations and various research studies funded by the National Institute on Drug Abuse. We have found that addiction is more than a medical issue. It encompasses physical, psychologic, and sociologic issues. The majority of women who are drug-dependent do not seek prenatal care and therefore have an increased incidence of medical and obstetrical complications. The common medical complications seen in these women include anemia, poor nutrition, and various infectious problems including sexually transmitted diseases, especially HIV disease. Many women also present with the complications associated with needle use, including hepatitis, thrombophlebitis, cellulitis, abscesses, and ulcers.

Obstetrical complications abound. These include amnionitis, chorioamnionitis, premature rupture of membranes, abruptio placentae, intrauterine growth retardation, and preterm birth. The host of medical and obstetrical complications experienced by the women demand that a large number of their infants be admitted to neonatal intensive care units for observation and treatment. Moreover, the women are further weakened by unemployment, illiteracy, homelessness, and legal issues. Many environmental variables have contributed to addictive behaviors. At least 70% of drug-dependent women have experienced sexual abuse before reaching age 16 years and 83% have a parent who was addicted to alcohol or illicit drugs (4).

Drug addiction is one of the most debilitating diseases due to its chronicity and intermittent relapses. Moreover, and specifically with regard to children, it has long-range and intergenerational implications.

Addiction is even more devastating to drug dependent women and mothers, who frequently do not have the resources for recovery available to them. Moreover, a host of barriers exist as a result of social welfare, medical, and legal systems. Substance abuse treatment services are insufficient and inadequate to meet the needs of pregnant women in this country and others. These needs would be better served if perinatologists and neonatologists would develop teams of professionals within existing services for high-risk pregnant women. Unfortunately, negative attitudes towards this population may generate resistance to treating them. Many in the medical profession ascribe to the outmoded concept that addiction is a problem of the criminal justice system and not in the realm of the medical clinician despite the fact it is clearly a chronic relapsing disease. The trend toward legal intervention in the pregnancies of women with alcohol and other drug problems discourages many from seeking care because they fear criminal prosecution or losing their children to foster care. Very often alcohol and other drug problems are not diagnosed because physicians and other clinicians lack adequate training in the identification of substance abuse and the consequences of drug use during pregnancy.

Additional treatment barriers for pregnant women who are substance abusers encompass the frequent naivete that exists concerning their needs as well as the milieu in which women exist, making it difficult for them to access traditional treatment services. Such issues include inadequate gender-sensitive treatment modalities; lack of cultural sensitivity in treatment programs; inattention to ethnic issues in treatment programs; nonexistence of a culturally and ethnically balanced staff; and the absence of

**Table 1.** Drug Abuse Treatment Schema for HIV Infected Drug Dependent Women

| | | | |
|---|---|---|---|
| **OUTREACH SERVICES** | | **BIOLOGICAL/ PHYSIOLOGICAL/MEDICAL** | |
| Community Liaison | Immediate Access to Treatment | For Potential Infections | Pharmacological |
| ◊ Indigenous workers | ◊ Provide transportation | ◊ Assess complications of drug abuse and HIV infection: tuberculosis, hepatitis, CMV | ◊ Methadone Maintenance |
| ◊ Mobile van | ◊ Coordinate intake | ◊ Assess various organ systems: hepatic, renal, cardiovascular pulmunary, GI, CNS | ◊ Psychotropic Medication |
| ◊ Distribution of prevention and educational information | ◊ Medical | ◊ Assess for STD's | ◊ Antibiotics for bacterial infections |
| ◊ Work with community organizations, churches recreational centers, etc. | ◊ Drug Abuse Treatment | ◊ Assess immunologic status | ◊ HIV drug treatment |
| | ◊ Psychiatric evaluation | | ◊ Clinical trials |
| **PSYCHOLOGICAL/BEHAVIORAL** | | **DEMOGRAPHIC/SOCIOCULTURAL** | |
| Life-Skills Management | Psychological | Survival Management | Sociological Consideration |
| ◊ Defining and accessing problems associated with addiction | ◊ Psychiatric assessment | ◊ Housing | ◊ Gender |
| ◊ Attutudes, beliefs, knowledge and expectation modification | ◊ HIV counseling | ◊ Clothing | ◊ Race |
| ◊ Problems solving | ◊ Nutritional counseling | ◊ Food | ◊ Social Class |
| ◊ Coping mechanisms | | ◊ Financial and Budgetary | ◊ Economics |
| ◊ Relapse prevention | | | ◊ Culture |
| ◊ Social-skills competence | | | |

STD -Sexually transmitted disease
GI - gastrointestinal
CMV- Cytomegalovirus
CNS- Central Nervous System

easily available transportation to treatment. Hospitals and communities must be aware of the need for outreach services for women not only with regard to prevention issues, but also with regard to access to treatment and follow-up.

Appropriate treatment services for these women, especially opiate-dependent individuals, has been established (12). Table 1 outlines the essentials, all of which are generally available in most medical centers. For maximum recovery rates, a dedicated clinician who realizes the need to coordinate such services for addicted women is needed. Since the medical needs of these women are so overwhelming, a perinatologist in conjunction with a neonatologist and psychiatrist should lead the team of professionals necessary to encompass, in addition to the physiological and psychological effects of substance abuse, the tremendous sociological issues that exist. Nearly 60% of addicted women have moderate to severe psychiatric diagnoses, frequently depression. Women will not recover if their co-morbidity issues are not identified and treated.

Research has shown that we can rehabilitate a significant number of women who have enrolled in comprehensive treatment services during pregnancy (3). We can reduce maternal and infant morbidity and mortality. When maternal medical and obstetrical complications are treated, a similar outcome has been seen in the drug-dependent mothers as in the drug-free mothers of the same socioeconomic and ethnic class. Moreover, one can reduce the incidence of low-birth- weight infants from nearly 50% to less than 20%, which is a significant reduction in terms of neonatal morbidity, mortality, and medical costs.

We must recognize addiction as a chronic, relapsing disease. Because each addicted woman is different from all others, treatment plans should be individualized. Comprehensive services must include high-risk prenatal care and clinics must be staffed by obstetricians specifically trained in the field of addiction and high-risk pregnancy. Additional treatment modalities should include individual, group, and family therapy.

The families of drug-addicted women have higher levels of family conflict and physical violence and lower levels of family cohesion. Therefore, treatment must respond to every one of the medical and social variables that complicate addiction and recovery. The women have problems associated with support issues, food, access to housing, and day care, which are issues that are most overwhelming to the recovering female addict. Relapse is imminent when daily survival is at risk.

AIDS prevention, counseling and testing, and educational services in the form of perinatal and parenting classes must be available. Services should be aimed at eliminating drug use, developing personnel resources, improving family and interpersonal relationships, reducing and eliminating socially destructive behavior, and facilitating obtainable adaptation for new parents within their environment.

We are fortunate to have a medication for the treatment of opiate dependence that, when coupled with the services described in Table 1 has the best success rate for recovery. Sadly, for opiate-addicted patients in general and for pregnant women in particular, there are many prejudices concerning the use of this safe and efficacious medication. Most of these attitudes stem from a lack of knowledge concerning the pharmacology and the appropriate prescribing instructions for methadone.

Methadone maintenance serves multiple purposes. Primarily, it attempts to remove the drug-addicted woman from the drug-seeking environment, eliminates the necessary illicit behavior, and prevents the "peaks and valleys" in the maternal heroin level that may occur throughout the day. Maternal nutrition is usually improved. The patients are more amenable to prenatal care and social/psychological rehabilitation. Since the drug-seeking behavior is eliminated, the women are more able to prepare for the birth of the children, begin homemaking, engage in counseling, and in many cases, complete their education and gain employment.

Medical withdrawal may be considered, but in only very few patients, and caution is warranted. The patient who before pregnancy has been maintained on a relatively high dose of methadone (>80 mg), the patient who in the hospital setting has needed a large amount of methadone for control of withdrawal symptoms, or the patient who insists that she deliver a drug-free infant may all require a slow dosage reduction regimen during the course of pregnancy.

Debate has existed for several decades regarding the efficacy of medically withdrawing a pregnant woman from methadone. To date, there is no evidence that methadone use has an adverse effect on the pregnant woman. On the contrary, methadone has been shown to "normalize" endocrine and immunological functions (13). Moreover, Zuspan et al. (14) studying the effects of withdrawal during the last half of pregnancy, found that lowering methadone levels could induce a marked fetal response of the adrenal gland, with increased amniotic fluid levels of epinephrine and a response of the other sympathetic nervous system components manifested by increasing amniotic fluid levels of norepinephrine. This response was blunted if the methadone dose was increased. They concluded that withdrawal from methadone late in pregnancy should not be recommended unless the fetus can be biochemically monitored for stress.

Moreover, the course of abstinence in heroin addicts during pregnancy is not smooth. Pregnant women medically withdrawn from narcotics have the same high rate of recidivism as nonpregnant addicts and thus become re-subjected to all of the attendant complications of illicit heroin use. The return to illicit heroin injection exposes the fetus to the risk of varying narcotic levels, which may well result in transient periods of fetal depression alternating with abstinence. Uterine contractions during maternal withdrawal cause an intermittent obstruction to placental perfusion resulting in intermittent fetal hypoxia.

The multiple medical and pharmacologic complications during parenteral heroin abuse must be considered before rejecting methadone maintenance treatment or recommending withdrawal from methadone during pregnancy. Also, polydrug or alcohol abuse may become an increasing problem if maintenance treatment is denied, if a dose reduction of methadone is carried out too rapidly, or if selected maintenance doses of methadone are too low to prevent supplementation with heroin. Many patients in these circumstances will use either illicitly obtained methadone or heroin. Others may turn to chronic abuse of tranquilizers, sedatives, barbiturates, or alcohol. Unfortunately, many well-meaning but poorly informed physicians may prescribe drugs such as diazepam for pregnant patients to blunt abstinence symptoms during methadone withdrawal. The use of such drugs not only may substitute another type of dependency, but also may be quite hazardous to the fetus.

It has been recommended that heroin-addicted women in an outpatient situation be given methadone maintenance using an adequate dose that still maintains the woman in a comfortable state (15-16). Depending on the duration of her addiction, the duration on methadone, her own metabolic needs, and concomitant medications such as Rifampin or Dilantin, an adequate dose that maintains the woman in homeostasis may vary from 35 to 80 mg. Each woman must have her dose individualized while appropriate monitoring of the pregnancy continues. With this therapy, injection of heroin is usually infrequent. It must be emphasized that informed consent for the use of methadone must be obtained, with all of the possible effects of methadone on both mother and baby being explained fully by the treating physician before initiating methadone therapy. Risks attendant to refusing treatment must also be fully explained.

Despite all the above justifications to use methadone in pregnant, opiate-addicted women since the benefits far override the risks, patient preference or relative dose may still indicate a need for dosage reduction or medical withdrawal. Although no scientific studies have determined a safe time for medical withdrawal, our clinical judgment recommends that it should be performed slowly, preferably decreasing the dose by 5 mg every 2 weeks. It is not advised before 14 weeks of gestation because of a potential risk of inducing abortion, and it should not be performed after the 32nd week of pregnancy because of possible withdrawal-induced fetal stress which may precipitate labor (17).

When withdrawing a pregnant woman from methadone, experienced individuals aware of the numerous perinatal nuances must be in charge. Ideally, the procedure should be performed in a perinatal unit with fetal monitoring so that, at the first sign of uterine irritability with the potential of the onset of labor, emergency measures can be taken and methadone resumed. The patient must be carefully selected, accounting for previous obstetrical history, length of addiction, and history of drug treatment failures (17).

Despite numerous attempts to withdraw some patients that requested it in Philadelphia, we had nearly a 100% failure rate for achieving a zero dose level. This was due either to the patient's inability to continue medical withdrawal (because of discomfort or the return to illicit drugs) or the perinatologist's advice because of obstetrical issues, particularly threatened onset of labor.

Some basic principles are important when managing pregnant, opiate-dependent women with methadone. The pharmacology of methadone in the pregnant woman has been well evaluated. Methadone readily crosses the placenta and has been identified and measured in amniotic fluid, cord blood, and neonatal urine (18-22). In the second trimester, both unchanged methadone and two major N-demethylated metabolites have been identified in amniotic fluid, while the cord blood levels were below the lower limits of detection. At delivery, the mean ratio of amniotic fluid methadone to maternal plasma methadone was approximately 0.73, and the ratio of cord blood to maternal plasma levels was approximately 0.5 (23-24).

Methadone is widely distributed throughout the body after oral ingestion, with extensive nonspecific tissue binding creating reservoirs that release unchanged methadone back into the blood, thus contributing to its long duration of action (25). After ingestion of a maintenance dose of methadone, peak plasma levels occur between 2 and 6 hours, with less than 6% of the ingested dose in the total blood volume at this time (26-28). Lower sustained plasma concentrations are present during the remainder of the 24-hour period. Methadone is metabolized primarily by the liver, and is excreted in the urine and feces both as unchanged methadone and a variety of metabolites. Studies of methadone in pregnant women show marked intra- and inter-individual variations, with a plasma level somewhat lower after a given dose during pregnancy than following delivery. This decrease in available methadone can be accounted for by an increased fluid space, a large tissue reservoir for storing methadone, and drug metabolism by both the placenta and the fetus (27). Methadone-maintained women frequently complain of increasing withdrawal as pregnancy progresses and frequently need elevations of their oral dose in order to maintain the same plasma level and remain withdrawal free (3).

Drug-dependent women in methadone programs before pregnancy can be maintained on their pre-pregnancy dosage when pregnancy is diagnosed, while women who are using street drugs should be admitted to the hospital for intensive medical and psychological evaluation and determination of an adequate methadone dosage. The heroin-addicted patient is initially evaluated for signs and symptoms of withdrawal, which include lacrimation, irritability, rhinorrhea, nausea, vomiting, abdominal cramping, uterine irritability, increased fetal activity, or rarely, hypotension. If evidence of withdrawal is present, 10 mg of methadone is given. Additional 5-mg doses are given every 4 to 6 hours if withdrawal symptoms are present. On the next day, the previous day's total dosage is given as the maintenance dose. The patient is evaluated for withdrawal, as previously, and supplemental doses are given as needed. Most patients are able to be well controlled on a dosage of between 20 and 35 mg daily after the first 48 to 72 hours of treatment. Subsequent to discharge, though, many patients will need a 10-to 30-mg increase during the pregnancy. The magnitude of the increase is dependent on the concomitant use of other drugs and continued use of heroin, as well as the potential changes in the third trimester. The latter, due to the altered methadone pharmacokinetics, may necessitate a dosage increase to produce a steady-state condition (29). Although the above have not been evaluated by a standard research design, nearly 1000 women have been placed on methadone using these conservative principles in our center without any untoward effects on their pregnancies. However, no studies have been done to verify its efficacy. Some advocate splitting the patient's dose (half in morning and half later in the day) in order to facilitate steady-state maintenance.

Since no adverse effects have been reported regarding maternal health during the pregnancy, many investigators have been advocates of restricting methadone to pregnant women on behalf of the fetus. Diverse findings between studies reflect the myriad of confounding variables that are present within human populations. The women differ on amounts of daily methadone dose, length of methadone maintenance during pregnancy, and the amount of prenatal care. A large percentage of pregnant women maintained on methadone continue to use a number of other drugs such as heroin, diazepam, cocaine, and barbiturates. In addition, 90% are moderate to heavy smokers, and alcohol consumption is quite prevalent (30-31).

One of the most emotional reactions to the use of methadone in pregnant women is the opinion that it increases the incidence and severity of neonatal abstinence. Therefore, some clinicians believe that since maternal heroin use may result in less neonatal abstinence, this illicit drug is preferred over the use of methadone. This is an uninformed and short-sighted approach to a complex issue.

It has been shown that 60% of infants exposed to methadone may have symptoms that require treatment (32). This issue has been the cause of an ongoing debate among neonatalogists in this field. Initial studies related the adverse effects of methadone with regard to the dose in the last trimester of pregnancy and its impact on neonatal abstinence (33). In comparing neonatal abstinence in heroin- with methadone-exposed infants, studies have had numerous design flaws including the inclusion of too few subjects, the use of noncomparable daily doses, and the failure to evaluate the results critically in light of infant maturity and other confounding variables.

Some studies have reported that the degree of neonatal withdrawal may be correlated with maternal methadone dose in the last trimester of pregnancy and a dosage decrease to less than 20 mg has been recommended. However, the relationship between maternal methadone dose and the presence of withdrawal symptoms has been difficult to establish. Although Ostrea et al. (15) and Madden et al. (16) reported a significant relationship between severity of withdrawal and methadone dose during pregnancy, other investigators (24) (34) (35) found no relationship between severity of withdrawal and maternal methadone dose. Kaltenbach et al. (36) examined maternal methadone dose and infant outcome for 147 women maintained on low dose (5 to 40 mg), moderate dose (41 to 60 mg), or high dose (>60 mg) during pregnancy. They found no differences among the groups in terms of birth weight, gestational age, or the number of days the infant required medication for abstinence. Multiple regression was used to determine factors predicting birth weight, gestational age, and days on medication for abstinence. There was no significant association for polydrug abuse, opiate use, use of other drugs other than opiates, average methadone dose, total months on methadone, or gender of the infant.

Although neonatal abstinence and the concomitant neurobehavioral characteristics can be directly attributed to in utero methadone exposure, further delineation of the effects of prenatal methadone exposure is a complex task. It should be emphasized that although infant developmental scores in a number of studies are lower when compared with nondrug-exposed infants, overall scores are within the normal range (31) (37) (39). Perinatal and developmental outcomes associated with prenatal methadone exposure must be viewed within a multifactorial perspective, i.e., the primary pharmacological-toxicological effects produced by the drug(s); postnatal environmental interactive effects; and possible genetic effects (40). Furthermore, it is essential that the use of methadone be viewed within the appropriate context. Methadone is a licit medication used to treat a chronic relapsing disease and, as with any medication, the risk/benefit ratio must be considered. When methadone maintenance is provided for pregnant drug-dependent women within a comprehensive treatment program that addresses the medical, obstetrical, addictive, and psychosocial issues, maternal and infant morbidity and mortality are reduced, and the developmental and cognitive functioning of the progeny are not impaired (36).

Infants undergoing abstinence symptoms can be easily evaluated using a scoring system (5). With appropriate assessment and treatment, neonatal methadone abstinence will result in moderate morbidity and no mortality. Moreover, with the greater chance of full-term birth in the methadone-exposed infant, the more severe sequelae of preterm birth are eliminated. Clearly, the benefits of methadone maintenance in the prenatal period far override the risks.

In order to improve the outcomes of addicted pregnant women and their children, clinicians must be cognizant of the many complicating issues and we must remove the barriers to effective prevention, intervention, and treatment for them. We must provide appropriate treatment by caring professionals who are knowledgeable in the field of substance abuse, and we must put the treatment of women into the context of their reality. For example, the woman who is drug dependent without any resources or support systems and who has three children cannot be asked to come every day to a treatment program. To be able to know what she must do for her children economically, emotionally, and spiritually, residential treatment should be available.

Aside from the intensive rehabilitation and medical treatment, these women need extensive educational and job training so that they can become productive citizens and loving mothers who will positively influence the development and socialization of their children. Residential treatment can serve to eliminate the medical and social problems experienced by the drug-dependent women and their children, particularly those who are cocaine-dependent and for whom there is no pharmacotherapeutic agent such as methadone available.

Incarceration is not the appropriate choice for drug-dependent women who have not had the financial, emotional, educational, or spiritual advantages that most citizens have had. Instead, treatment on need must be available for the mother with her children, not only for the rehabilitation of these women and children, but also for the sake of future children. Appropriate services for the follow-up of pregnant drug-dependent women are essential. The issues of child abandonment and increased needs for foster care must also be added and planned for appropriately.

In summary, maternal drug-abuse and its potential effects on neonatal development are major problems that continue to perplex and challenge the best minds in medicine and government. While we have learned a great deal, much remains to be understood. It is a dilemma that has yet to be appropriately addressed for a number of reasons. First, clinical studies have produced conflicting data because of confounding variables and poor methodology. Second, perinatologists, neonatologists, and primary-care physicians have shown little interest in providing clinical care for pregnant, substance-abusing women and their drug-exposed infants, or in conducting research related to neonatal development following in utero drug exposure. Third, there has been a distressing lack of education among both physicians and the lay public regarding the effects of licit and illicit drugs and the treatment of substance abuse. Few medical schools address these subjects. Fourth, government and private funding for research into issues of substance abuse and treatment needs an even greater focus and commitment. The above treatment and research recommendations are the least that we must do for the health of addicted women and to ensure that their future children will not be tormented by drug addiction and the likelihood of being physically and psychologically disabled throughout their lives (41).

## References

1. Finnegan L P. Influence of maternal drug dependence on the newborn. In: Toxicologic and Pharmacologic Principles in Pediatrics, edited by Kacew S, Lock S. Hemisphere Publishing Corp., New York, N.Y., 1988, pp 183-198.
2. Kaltenbach K, Finnegan L P. Prenatal narcotic exposure: Perinatal and developmental effects. Neurotoxicology 10:597-604, 1989.
3. Finnegan L P, Wapner R J. Narcotic addiction in pregnancy. In: Drug Use in Pregnancy, edited by Neibyl J R. Lea & Febiger, Philadelphia, PA, 1987, pp 203-222.
4. Regan D O, Ehrlich S M, Finnegan L P. Infants of drug addicts: At risk for child abuse, neglect,

and placement in foster care. Neurotoxicol Teratol 4:315-319, 1987.

5. Finnegan L P. Neonatal abstinence syndrome: Assessment and pharmacotherapy. In: Neonatal Therapy: An Update, edited by Rubaltelli F F, Granati B. Elsevier, New York, N.Y., 1986, pp 122-146.
6. Pasto M E, Graziani I J, Tunis S L, et al. . Ventricular configuration and cerebral growth in infants born to drug dependent mothers. Pediatr Radiol 15:77-81, 1985.
7. Ryan L, Ehrlich S, Finnegan L P. Cocaine abuse in pregnancy: Effects on the fetus and the newborn. Neurotoxicol Teratol 9:295-299, 1987.
8. Finnegan L P, Mellot J M, Ryan L M, et al. . Perinatal exposure to cocaine: Human studies. In: Cocaine: Pharmacology, Physiology and Clinical Strategies, edited by Lakoski J M, Galloway P, White F J. CRC Press, New York, N.Y., 1991,
9. Falloon J, Edy J, Winer L, et al. . Human immunodeficiency virus infection in children. J Pediatr 114:1-30, 1989.
10. Johnson E M. Testimony of Elaine M. Johnson, Ph.D. Associate Administrator for Prevention and Director of the Office of Substance Abuse Prevention in the Alcohol, Drug Abuse and Mental Health Administration, before the Subcommittee on Health and the Environment of the US House Committee on Energy and Commerce, 1990.
11. Brien J F, Clarke D W. Disposition and fetal effects of ethanol during pregnancy. In: Toxicologic and Pharmacologic Principles in Pediatrics, edited by Kacew S, Lock S. Hemisphere Publishing Corp., New York, N.Y., 1988, pp 199-222.
12. Finnegan L P, Hagan T, Kaltenbach K. Opioid dependence: Scientific foundations of clinical practice. Pregnancy and substance abuse: Perspective and directions. Bull N Y Acad Med 67:223-239, 1991.
13. Kreek M J. Methadone in treatment: Psychological and pharmacological issues. In: Handbook on Drug Abuse, edited by Dupont R I, Goldstein A, O'Donnell J. NIDA U.S. Department of Health and Human Services, Rockville, MD, 1979,
14. Zuspan F P, Gumpel J A, Mejia-Zelaya A. Fetal stress from methadone withdrawal. Am J Obstet Gynecol 122:43, 1975.
15. Ostrea E M, Chavez C J, Strauss M E. A study of factors that influence the severity of neonatal narcotic withdrawal. J Pediatr 88:542-545, 1976.
16. Madden J D, Chappel J M, Zuspan F, et al. . Observation and treatment of neonatal narcotic withdrawal. Am J Obstet Gynecol 127:199, 1977.
17. Finnegan L P. Treatment issues for opioid dependent women during the perinatal period. J Psychoactive Drugs 23:191-201, 1991.
18. Kreek M J. Medical safety and side effects of methadone in tolerant individuals. J Am Med Assoc 223:665, 1974.
19. Peters M A, Tunrbow M, Buchenauer D. The distribution of methadone in the nonpregnant, and fetal rat after acute methadone treatment. J Pharmacol Exp Ther 181:273, 1972.
20. Sanner J H, Woods L A. Comparative distribution of tritium labelled dihydromorphine between maternal and fetal rats. J Pharmacol Exp Ther 148:176, 1965.
21. Yeh S Y, Woods L A. Maternal and fetal distribution of H3 dihydromorphine in the tolerant and non-tolerant rat. J Pharmacol Exp Ther 174:9, 1983.
22. Blinick G, Inturrisi C E, Jerez E, et al. . Amniotic fluid methadone in women maintained on methadone. Mt Sinai J Med 41:254, 1974.
23. Blinick G, Inturrisi C E, Jerez E, et al. . Methadone assays in pregnant women and progeny. Am J Obstet Gynecol 121:617, 1974.
24. Rosen T S, Pippenger C E. Disposition of methadone and its relationship to severity of withdrawal in the newborn. Addictive Dis 2:169, 1975.
25. Dole V P, Kreek M J. Methadone plasma level: Sustained by a reservoir of drug in tissue. Proc Natl Acad Sci 70:10, 1991.
26. Inturrisi C E, Verebely K. A gas liquid chromatographic method for the quantitative

determination of methadone in human plasma and urine. Chromatography 65:361, 1972.
27. Kreek M J, Schecter A, Gutjahar C L, et al. . Analyses of methadone and other drugs in maternal and neonatal body fluids: Use in evaluation of symptoms in a neonate of mother maintained on methadone. Am J Drug Alcohol Abuse 1:409, 1974.
28. Sullivan H R, Blake D A. Quantitative determination of methadone concentration in human blood, plasma, and urine by gas chromatography. Res Commun Chem Pathol Pharmacol 3:467, 1972.
29. Kreek M J. Opioid disposition and effects during chronic exposure in the perinatal period in man. In: Advances in Alcohol and Substance Abuse. Parts 3 and 4, edited by Stimmel B. Halworth Press Inc., New York, NY, 1982,
30. Rosen T S, Johnson H L. Children of methadone maintained mothers: Follow-up to 18 months of age. J Pediatr 101:192-196, 1982.
31. Wilson G S, Desmond M M, Wait R B. Follow-up of methadone treated women and their infants. Health, development and social implications. J Pediatr 98:716-722, 1981.
32. Finnegan L P, Ehrlich S M. Maternal drug abuse during pregnancy: Evaluation of pharmacotherapy for neonatal abstinence. In: Modern Methods of Pharmacology Testing and Evaluation of Drugs of Abuse. Vol. 6 , Wiley-Liss Inc., New York, NY, 1990, pp 255-263.
33. Harper R, Solish G, Feingold E, et al. . Maternal ingested methadone, body fluid methadone, and the neonatal withdrawal syndrome. Am Inst Obstet Gynecol 129:417, 1977.
34. Blinick G, Jerez E, Wallach R C. Methadone maintenance, pregnancy and progeny. J Am Med Assoc 225:477, 1975.
35. Stimmel B, Goldberg J, Murphy R, et al. . Fetal outcome in narcotic dependent women: The importance of the type of maternal narcotic used. Am J Drug Alcohol Abuse 9:373-395, 1983.
36. Kaltenbach K, Finnegan L P. Methadone maintenance during pregnancy: Implications for perinatal and developmental outcome. In: Perinatal Substance Abuse: Research Findings and Clinical Implications, edited by Sonderegger T B. The Johns Hopkins University Press, Baltimore, MD, 1992, pp 239-253.
37. Chasnoff I J, Schnoll S H, Burns W J, et al. . Maternal nonnarcotic substance abuse during pregnancy: Effects on infant development. Neurobehav Toxicol Teratol 6:277-280, 1984.
38. Kaltenbach K, Finnegan L P. Perinatal developmental outcome of infants exposed to methadone in utero. Neurotoxicol Teratol 9:311-313, 1987.
39. Strauss M E, Lesson-Firestone J K, Chavez C J, et al. . Behavioral concomitants of prenatal addiction to narcotics. J Pediatr 89:842-846, 1976.
40. Falek A, Madden J J, Schafer D A, et al. . Individual differences in opiate-induced alterations at the cytogenetic, DNA repair, and immunological levels: Opportunity for genetic assessment. In: Genetic and Biological Markers in Drug Abuse and Alcoholism, edited by Braude M C, Chao H M. NIDA Department of Health and Human Services, Rockville, MD, 1986,
41. Finnegan L P. Perinatal substance abuse: Comments and perspectives. Seminars in Perinatology 15(4):331-339, 1991.

# 17

# Tuberculosis. The Comeback of a Killer

**J. H. Lowinson, M. Gourevitch**
Albert Einstein College of Medicine of Yeshiva University, Division of Substance Abuse, Bronx, New York, NY, U.S.A.

## Summary

Tuberculosis, the single largest cause of death worldwide in the 19th and first half of the 20th century by an infectious agent, was considered a controllable disease in the 1950's after the discovery of streptomycin, PAS and isoniazid. The incidence of tuberculosis began to decline in the United States and in Western Europe in the 1960's and 1970's resulting in the closing of sanatoria and clinics devoted to the treatment of this dread disease. This paper will deal with some of the myths related to TB including the one that, like small pox, for which a vaccine had been developed, TB could be totally eradicated. In the late 1970's and the 1980's, a gradual increase in TB began to re-emerge and reached epidemic proportions in the late 1980's in New York City and other urban areas of the United States. This upsurge was alarming and mysterious. In 1990, the WHO convened a meeting in Geneva where the conclusions drawn by representatives of TB programs from all over the world were alarming: 2.9 million deaths were caused by tuberculosis in 1990, "making this disease the largest cause of death from a single pathogen in the world". In June, 1992, the WHO issued a press release about the rise of tuberculosis in industrialized countries including the United Kingdom and Italy in addition to the United States. It seemed evident that there must be a common factor. Although socioeconomic factors such as crowding, homelessness and poor nutrition were contributing factors, the main factor for this global threat was the human immunodeficiency virus (HIV).

In consideration of the large number of intravenous drug users who have become HIV+, (many of whom have developed AIDS), this paper will deal with the role that substance abuse treatment programs (particularly Methadone Maintenance Programs) are playing in the identification, treatment and prophylaxis of tuberculosis. Patients are being tested and identified and environmental protection is being introduced into the clinics. While these programs cannot address the socioeconomic factors that have always played a role in the transmission of tuberculosis, efforts are made to improve the living conditions of the patients enrolled in their clinics. A major factor that has enhanced the seriousness of the medical aspects of tuberculosis today is the emergence of multidrug resistant (MDR) TB. Although non-compliance has contributed to the development of MDR TB, it has long been recognized that the drugs most effective in treating TB result in resistance even when taken as directed. To the extent that time permits this paper will attempt to deal with the above issues and the efforts being made to reduce the risks inherent in a world where co-existing tuberculosis and HIV are so prevalent.

Infectious diseases have been and remain the leading cause of death world wide, and tuberculosis heads the list as the single pathogen responsible for the largest number of deaths. It is responsible for more deaths world wide than cancer or cardiovascular disease. Annually there are eight million new cases of tuberculosis, and 2.9 million deaths. The greatest killer in history, tuberculosis has claimed more than a billion lives world wide. One third of the world's population harbors the tubercle bacillus, and in some developing countries as many as 67% of the deaths are caused by tuberculosis.

From the earliest days of recorded medical history going back to Hippocrates, tuberculosis has been the most dreaded scourge. Many of the most creative people, such as Keats, Byron and Chopin among others, have fallen victim to this disease at a very young age and at the height of their creative talents (1).

As AIDS is now, it was feared by people of all classes and eluded the efforts of the best scientists to find an effective treatment.

Today we are confronted with the deadly dyad of HIV and TB. The synergism of these two pathogens has created a public health problem that threatens every segment of our society.

DOUBLE TROUBLE... This is the issue with which we are concerned here today, but first let's briefly review the history of tuberculosis.

In general, tuberculosis was viewed as a weakness of the spirit. In 16th century Italy, Fracastora included phthisis (as tuberculosis was then known and which is the Greek word for wasting) in a work on contagion. Then the Decree of Lucca, promulgated in 1699, required physicians to notify the General Sanitary Council of the names of patients with phthisis. And it was Villemin in 1865 who formally demonstrated that TB was contagious by infecting rabbits with pus from a human tubercle. In 1882 Robert Koch, a German country doctor, shook up the medical world with his isolation of bacterial rods from "tubercles" or freshly developed abscesses, from human beings who had died from tuberculosis. With this discovery, it was finally established that TB was in fact a bacterial disease (2).

There ensued a feverish search in Europe and America to discover a cure for this disease which killed one in seven people world wide. Decades would pass before Albert Shatz, a young assistant in Selman Waksman's laboratory at Rutgers University would discover in 1943 that streptomycin was effective in inhibiting in vitro the growth of the Mycobacterium bacillus (3).

In 1944 the first trial of parenteral streptomycin in human tubercular patients led to a dramatic improvement of symptoms even though sputa failed to turn negative and resistance developed in most cases. (Adding PAS to the medical regimen for a period of 4 months did result in negative sputa).

In 1952 the first trial of isoniazid taken orally appeared promising. Symptoms cleared but in only 8 of 44 patients did sputa turn negative. Moreover, resistance developed frequently and rapidly. It soon became evident that no single drug in the pharmacopoeia was a miracle drug capable of eradicating a patient's infection--and the same holds true today. A combination of drugs was recognized as necessary.

In 1963 Rifampin was added to the armamentarium of anti-tuberculosis medications. As powerful as isoniazid and less toxic than streptomycin, it could also be taken orally. Further studies showed that a combination of INH and Rifampin yielded even better results than either taken separately; with the introduction of additional anti-TB drugs in the 1960's (e.g. ethambutol, pyrazinamide and others), the hope (and illusion) that tuberculosis could be eradicated led to a spirit of optimism (4). In the United States, a decline in the incidence of tuberculosis had actually begun well before the discovery of chemotherapeutic agents.

Improved living conditions, better housing and less crowding seemed to be factors contributing to this reduction in the number of active cases. As a result of this dramatic decline in the US and in other industrial nations, the need for sanatoria and special hospital beds and clinics was questioned and they were closed. Even the sanatorium in Davos made famous by Thomas Mann in The Magic Mountain was closed and faded into history. Many experts in the field questioned the wisdom of these moves and expressed the idea that, although tuberculosis appeared to be waning, we could expect a resurgence of this deadly disease.

Beginning in the late 1970's, there was a gradual increase in the number of cases of tuberculosis in the United States and strains resistant to more than one chemotherapeutic agent began to emerge with increasing frequency. Pulmonary physicians as early as 1978 (Lee Reichman et al.) warned that tuberculosis was on the increase but their words fell on deaf ears. In reality, although the incidence of tuberculosis had dramatically diminished by the 1960's, it had never disappeared (5).

In 1990, articles in the daily newspapers reflected the recognition by the news media that there was indeed a deadly strain of tuberculosis reaching epidemic proportions in New York City. That same year the World Health Organization convened a summit meeting in Geneva to address the growing number of TB cases in developing nations, especially the subSaharan countries. There was a widespread fear that Africa was lost. In 1992, the WHO added at least 10 European countries (including the Netherlands, Sweden, Finland and Italy) to the list of those regions where TB had re-emerged. Although the increase was now observed to be primarily associated with the HIV epidemic, there was general recognition that poverty and a growing homeless population were also significant contributory factors.

In New York City in 1991, the number of tubercular cases continued to rise. Case rates which had been rising since 1979, were then 50.2 per 100,000 population. There were 3,673 new cases of tuberculosis that year. The case rate in 1991 was the highest in two decades, having risen 143% since 1980. One third of all NYC cases reported in 1991 were resistant to two or more drugs, and nearly one in five were resistant to both INH and Rifampin, the two most effective oral anti-TB drugs.

Patients with a previous history of treatment for TB were more likely to have resistant strains than those who had never been treated: 44% had strains resistant to one or more anti-TB drugs, 36% had strains resistant to INH and 30% demonstrated resistance to both INH and Rifampin (6).

The fatality rate for multiple drug resistant MDR-TB in some series is between 40% and 60% in HIV- individuals and up to 80% in HIV+s. In 1990 and 1991, 3 outbreaks of MDR-TB in NYC hospitals were investigated by the CDC and the NYC Department of Health. Eighty-two (82) cases of MDR-TB were identified as probably occurring due to transmission within these institutions. At least 85% of the patients were HIV positive.

The pool of prevalent latent infections is a major source of new cases occurring in individuals whose immune systems can no longer contain their infection. The extent to which recently acquired infection contributes to the incidence of new cases is uncertain, but may be greater than previously recognized.

Prior to the era of HIV, it was known that the risk of progression from infection to disease is greatest in the two years following infection and tends to decline with the passage of time. The lifetime risk of disease is about one in ten unless preventive therapy is given, and prophylaxis with isoniazid is about 90% effective in preventing the disease. However, the course of TB in HIV+ individuals is dramatically different. Tuberculosis is an AIDS-defining disease in HIV infected patients and is often the first indication of HIV infection. While the lifetime risk of developing tuberculosis in the average healthy person is 10% over a lifetime, it is 8% a year in an HIV+ individual. In view of the fact that cure rates can be as high for HIV+ patients as for those who are HIV-, it is essential to start drug treatment without delay. Recognizing the synergism between tuberculosis and HIV infection, and the high percentage of co-infection among methadone patients, all methadone patients must be evaluated for tuberculosis infection and disease.

Although not all patients enrolled in methadone treatment are at equal risk for acquiring infection with HIV or tuberculosis, all patients should be screened for tuberculosis infection and disease. The Mantoux skin test is used to detect tuberculosis infection. A reaction of 5 mm or more is considered positive in HIV infected individuals; 10 mm or greater induration is considered positive in patients with a history of intravenous drug use and other risks for TB infection, and 15 mm or more is considered positive for patients without other risk factors for tuberculosis infection (7).

All patients except those who tested positive for tuberculosis infection in the past should undergo testing upon admission to the program and annually or semi-annually thereafter. HIV-infected drug users with cutaneous anergy have been shown to be at high risk of developing tuberculosis. Anergy testing is therefore recommended for HIV+ and HIV-serostatus unknown patients in methadone treatment. At Montefiore, anergy tests are placed simultaneously with the PPD for all patients regardless of HIV serologic status to eliminate the need for additional patient visits (8-9). A negative PPD together with a reactive anergy panel make the diagnosis of tuberculosis infection unlikely. It is important to point out that skin testing identifies tuberculosis infection. It does not diagnose active tuberculosis which requires radiologic tests and culture of the bacillus from the patient. Therefore all patients with reactive skin tests or who are HIV+ or HIV-unknown and anergic warrant evaluation for active disease with a chest xray.

The classical radiologic appearance of pulmonary TB is of the upper lobe infiltration with cavitation. In HIV+ patients the xray pattern is frequently atypical; infiltrates in any lung field may be found, and cavitation is uncommon. All too often this has led to a delay in the diagnosis of TB in HIV infected patients. Culturing M. TB sputum specimens can be a slow and tedious process, usually taking 3 to 8 weeks. Subsequently, determining susceptibility to a wide range of anti-TB drugs can be equally daunting. When in vitro sensitivity is not known, which is generally the case at the outset of therapy, treatment of patients with active TB should be started in a hospital on a 4 drug regimen, consisting of INH, Rifampin, pyrazinamide and streptomycin or ethambutol. This regimen is usually effective even against INH resistant strains and can be administered twice a week after an eight-week induction period of daily therapy. Sputum conversion is achieved more rapidly with this 4 drug regimen than with a 3-drug regimen of INH, Rifampin and PZA, and a patient who is treated with the 4 drug regimen is more likely to be cured and not relapse when compared with a patient treated for the same length of time with the

3 drug regimen.

It is well known that Rifampin decreases serum levels of methadone when the two medicines are taken concurrently. In this age of epidemic tuberculosis, with drug users bearing a disproportionate burden of active disease, more and more methadone-maintained patients are being started on Rifampin.

It is imperative that clinicians be prepared to work closely with methadone-maintained patients being started on potentially life saving medication. Patients must be advised in advance of the probable decrease in efficacy of their current methadone dose, and reassured that the clinician will raise the patient's methadone dose as rapidly as the patient feels it is necessary.

Only in this manner will patients be able to tolerate this critical drug, thereby increasing the likelihood that they will be compliant with the regimen and that they will complete their course of therapy. The risk of progression to disease in newly infected individuals is greatest in the two years following infection and tends to decline with the passage of time.

The risk of infection is about one in ten unless preventive therapy is given, and the successful results of prophylaxis is high. The drug of choice for prophylaxis is isoniazid given daily in a dose of 300 mg. Unlike treatment of active disease, isoniazid ideally should be given daily for prevention as there are no good studies showing effective prophylaxis unless the medication is taken daily. The most effective treatment outcome is associated with 300 mgs given daily under observation, and that is the regimen we try to follow at Einstein and Montefiore. However, because of fiscal and other constraints some groups give isoniazid two or three times weekly in appropriately higher doses.

Regardless of the regimen, the medication should be taken under direct observation. The reduction in the number of visits per week makes it possible for the staff to observe the administration of the medication. Otherwise it may be necessary to find a family member who will supervise the patient.

One of the greatest challenges in controlling tuberculosis is achieving compliance. Compliance can be increased by administering the anti-TB drugs when patients come in for their methadone, offering an opportunity for directly observed therapy (DOT) and directly observed prophylactic therapy (DOPT). This gives staff the opportunity to educate patients about the personal and public health benefits of taking the medication as prescribed, and to motivate the patient.

Methadone maintenance treatment clinics are ideal sites to accomplish multiple goals. When this treatment modality was originally developed by Dole and Nyswander in the 1960's, their main purpose was the rehabilitation of patients, to help them develop or rebuild normal lives. In the past three decades the patients have presented with increasingly difficult problems and now the clinics play a crucial role in the treatment of HIV infection, AIDS, and tuberculosis. Unlike AIDS for which there is no known cure at the present, tuberculosis is not only treatable but it is also preventable in a large percentage of these who are infected. One of the greatest challenges in controlling tuberculosis is achieving compliance.

Methadone maintenance clinics can play a crucial role in achieving compliance and other goals essential to a successful outcome.

Given the known synergism between HIV and TB, and the high percentage of HIV infection among many of the patients who are former intravenous drug users, the guidelines which have been issued by the CDC can be implemented in methadone clinics.

Compliance can be achieved by administering the anti-TB drugs when patients come in for their methadone offering an opportunity for directly observed therapy (DOT) and directly observed prophylactic therapy (DOPT). They also give staff the opportunity to educate patients about the personal and public health benefits of their taking the medication as prescribed, and to motivate the patients (10).

Treatment of subclinical infections in the general population is critical since 5 to 10% will develop active disease over their lifetime. The development of disease is significantly greater in HIV infected patients--and the greatest risk among this group seems to be in present (and past) intravenous drug users. HIV-positive patients progress from the infected to the active state at the rate of 8% per year and at a more rapid rate than the otherwise healthy individual. If treatment is started early enough in an HIV+ patient, the cure rate is usually high.

Van Etten History.

It might be of some interest to review the first pilot program where methadone was used to treat tubercular addicts in 1966 at Einstein in the Van Etten Hospital for Chest Diseases. Prior to the introduction of methadone, 95% of these patients signed out A.M.A. One year after the pilot was started, the figure was reversed; not only did the patients remain in the hospital until their sputa turned negative, but they returned to the hospital after discharge where they received their anti-TB medications and methadone. The irony of this successful experiment is that the results contributed to the closing down of special beds for tubercular patients.

The resurgent tuberculosis epidemic reflects a failure to maintain a public health infrastructure focused on treating active tuberculous patients until cured and on preventing active disease in high risk populations. Directing public health resources into targeted programs that enhance compliance with treatment regimens and expanding chemoprophylaxis among HIV+ individuals is not only possible but essential.

To achieve success in controlling tuberculosis, a sustained commitment to fund an adequate public health infrastructure is needed. There are multiple problems which include the need for better and more rapid diagnostic techniques to determine drug susceptibility, improved social conditions with the elimination of homelessness and crowded living conditions, the introduction of environmental controls especially in long term living facilities such as prisons and nursing homes, and the ability to ensure that patients complete their course of therapy.

Sheldon Landsman expressed it very effectively in a recent issue of the American Journal of Public Health when he suggested that the tuberculosis epidemic in New York City can be seen "as the superimposition of a biologic process--HIV infection-- upon a social and political landscape of inadequate policies, planning and funding (11)." The large number of HIV+ patients with active TB has increased the caseload of patients requiring active supervision, but there has not been a concomitant increase in programs or personnel to provide the supervision. And completion of therapy is the first and central priority of tuberculosis control.

## References


1. Abbott E C. Composers and Tuberculosis: The Effects on Creativity. CMA Journal 126:534-544, 1982.
2. Burke R. An Historical Chronology of Tuberculosis, Charles C. Thomas Publishers, Springfield, Illinois, USA, 1955.
3. Ryan F. The Forgotten Plague, Little, Brown & Co., Boston, 1993.
4. Jain A, Mehta V, Kulshrestha S. Effect of Pyrazinamide on Rifampicin Kinetics in Patients. Tubercle and Lung Disease 74:87-90, 1993.
5. Reichman L, Felton C, Edsall J. Drug dependence, a possible new risk factor for tuberculosis disease. Arch Intern Med 139:337-339, 1978.
6. Iseman M. Drug-resistant Tuberculosis. Clinics in Chest Medicine 10(3):341-353, 1989.
7. Barnes P, Bloch A, Davidson P, Snider D. Tuberculosis in Patients with Human Immunodeficiency Virus Infection. The New England Journal of Medicine 324:1644-1650, 1991.
8. Morbidity and Mortality Weekly Report, April 26 Vol 40 No RR-5, 1991.
9. Selwyn P, Hartel D, Lewis V, Schoenbaum E, Vermund S, Klein R, Walker A, Friedland G. A Prospective Study of the Risk of Tuberculosis Among Intravenous Drug Users with Human Immunodeficiency Virus Infection. The New England Journal of Medicine 320 (9):545-550, 1989.
10. Gourevitch M. Successful Directly Observed Chemoprophylaxis (DOC) and Therapy (DOT) of Tuberculosis (TB) in HIV-Seropositive (HIV+) and HIV-Seronegative (HIV-) Injecting Drug Users (IDU's) in a Methadone Maintenance Treatment Program (MMTP). In: Abstract presented at the APHA 121st Annual Meeting, San Francisco, 1993,
11. Landesman S. Commentary: Tuberculosis in New York City - The Consequences and Lessons of Failure. Am J Pub Health 83 (5):766-768, 1993.

# 18

# Lowering the Threshold: A Comparative Overview of AIDS Prevention Among Injecting Drug Users Using Outreach Technology

**C. D. Kaplan**
Institute for Psychosocial and Socioecological Research, University of Limburg, Maastricht, The Netherlands

## Summary

This paper provides an overview of AIDS prevention among drug users in the European and American context. The most recent epidemiological data in Europe and America indicate that drug users are the group greatest at risk to contract the HIV virus. Accumulating evidence shows that spread if HIV virus is associated with specific socio-ecological units such as neighbourhoods and social networks. These data support the idea that the prevention of the spread of the HIV in the drug users population should be focussed on specific geographical areas and on social networks. Outreach technology has been developed involving the use of indigenous drug users working in their own neighbourhoods and within their own networks together with professional health promotion workers. The technology of outreach is also associated with the concept of self organization drawn from the ecological sciences. This technology aims at lowering the threshold between relevant services and information and the drug user population. To illustrate this approach data is presented from the 'Boule de Neige' intervention project in the Euro-region of Germany, The Netherlands, Belgium and the Chicago Outreach Intervention Project.

In the World Health Organization (WHO) progress report issued in 1988 by the Global Program on Acquired Immunodeficiency Syndrome (AIDS) a simple truth was stated: " ... transmission requires the active participation of two persons." (1) AIDS is a communicable disease and in this way it is similar to the disease of drug addiction. Both the sharing of bodies and the sharing of needles are the major routes of transmission of AIDS. Both involve active social behavior. But there is more to this essential social dimension than two person relationships. The actual spread of AIDS in the community is more linked to the characteristics of the personal social networks in which the individual at risk participates where partners are selected and needles are obtained. For example, in the Boule de Neige (i.e. Snowball) project, an outreach project for the prevention of AIDS among drug users in the Euregio formed by the intersection of The Netherlands, Belgium and Germany, the credo for prevention has been "AIDS travels the same route as information, i.e. from friend to friend"(2). Thus, basic to the outreach technology reviewed in this paper is the recognition that the personal social network of the drug user, both for the spread of AIDS and its prevention, is essential for intervention. Without this recognition, a pessimistic conclusion seems evident when considering drug users. For instance, in the United States of America (USA), the statistics are bleak. Injecting drug user (IDU)-associated AIDS increased from 33.3% to 34.6% in a single year in 1991. In this rise is included 11.5% who were heterosexual partners of IDUs and 2.3% babies from IDU mothers or sexual partners (3). From 38 countries representing 60% of the global population who have reported data there were estimated 3.2 million IDUs representing a rate of 101/100.000 population. Extrapolating to the total population, the global figure could be as high as 5.4 million. With rates like this, Martin Plant at the 1993 Addiction Forum in Durham, England, cited Jonathan Mann, former director of the WHO Global Program, stating that in just six years the WHO goal of "Health for All by the Year 2000" might be confronted with quite a different reality - a world of accelerating misery, sick on AIDS spreading from IDUs to their sexual partners and their babies.

Turning the situation around, an optimistic conclusion begins with the recognition that the spread of AIDS might well be contained and even stamped out if prevention efforts are specifically targeted to sociological units such as personal networks. The USA data clearly indicate that despite general growth, AIDS varies considerably in the dynamics of its spread. Rates vary by group; the Black American rate is 11.9 times higher than White American rate and twice the rate of Hispanic Americans. The rate in the Northeast of the USA is 8.6 times higher than the Midwest. These kind of variations can also be seen in Europe. In Italy, for example, in a sample of 1500 IDUs attending drug clinics between 1985 - 1991 large variations in attack rates could be observed: 4.9%/year for Rome, 3.1% for Milan and 0.0% for Naples. Generally, prevalence rates have also been declining in Italy. Since 1987, the rate has dropped under the 5% mark. This suggests that in areas where prevalence was low in the 1980s further spread can be stopped.

Not only large ecological unities such as regions and cities show significant variations in AIDS indications. Neighborhoods and social networks can also be significant. For example, in Arnhem, a small Dutch city bordering on Germany, HIV prevalence in a methadone maintenance program sample of 135 was estimated to be 1.3% despite high levels of risk behavior. Further analysis showed that this anomaly could be explained by the fact that relatively few of the sample had contact with Amsterdam. In earlier prevalence studies being a German heroin tourist in Amsterdam was a main determinant of HIV risk (4). In contrast, not being a heroin tourist in Arnhem might function as a protective factor. A related finding can be seen in the USA National Research Council's study "The Social Impact of Aids". In New York City two epidemics could be distinguished. Each epidemic was characterized by being clustered in distinct neighborhoods and was settling into spatially and socially isolated groups. The report sounded an optimistic note that by targeted, intense and focused interventions in which the neighborhood is the key for prevention, positive outcomes could be anticipated.

**Outreach: Lowering the Threshold of Optimism**

From a social system point of view, positive changes in AIDS prevention among IDUs has been demonstrated in numerous studies cross-culturally. Exposure to drug abuse treatment as well as research itself is a significant factor. But, we also know from the international data that a dismally low percentage of the total population at risk ever is in the position to benefit from this exposure. This has stimulated the development of outreach technology to lower the threshold of exposure to treatment services while, at the same time, actively participating and intervening in the social networks of drug users. The impact of outreach technology should not be accessed solely on the basis of the objective of encouraging recruitment for treatment and health education. More essential in terms of long-term outcomes is the enablement of what, Sam Friedman, has called the "mobilization of subcultural change" (5). Outreach technology provides an instrument of effecting the social norms of IDUs that increases the likelihood of personal behavioral change. To be more specific, comparative studies of risk factors in the culturally wide variations of New York and Bangkok found that 80% of New York and 92% Bangkok IDUs reported changing behavior concerning AIDS. A common protective factor in both cities was talking about AIDS with drug using friends. Risk reduction is also associated in both cities with undergoing experiences that personalized the threat of AIDS such as counseling or testing. These user studies find risk reduction to be a social process involving discussions among drug using peers and with helping professionals that further personalize the threat. Within this social process the mechanism of subcultural change can be specified - new norms emerge and new rituals of discussion and counseling are added to the existing IDU normative framework.

Ernst Buning from the Amsterdam Municipal Health Department was asked by WHO Regional Office to inventorize outreach activities for drug users in light of the AIDS epidemic (6). The credo of most outreach teams is "don't wait for them to come to you, you go for them" and "make sure that you see and are being seen." In summary, Buning concludes that outreach projects have four important functions: contacting drug users who are not in touch with existing facilities; promoting information (safe drug use, safe sex), survival help and AIDS assistance; linking active drug users to drug help and treatment by referral and informative provisions; detecting new trends, groups and needs, i.e. intelligence. These functions can stimulate and reinforce the positive social processes of subcultural changes leading not only to risk reduction in behavior, but also virtually eliminating seroconversion in the networks of IDUs in which the outreach teams operate.

Perhaps the best example of the effectivity of outreach are the results of the Community Outreach Intervention Project in the School of Public Health, University of Illinois, Chicago. The results first reported by Wayne Wiebel, the project director and Mary Utne O'Brien in Paris in May, 1993 at the Third International Seminar on Drug Abuse, AIDS and Cities: New Protocols of Prevention and the next month in Berlin at the Ninth International Conference on AIDS provides the conclusive evidence. The project targeted IDUs not in treatment estimated at 90% of the Chicago IDU population. Networks, not individuals, were primarily targeted. The project employed workers recruited from the communities they targeted and were former addicts themselves. This methodology was termed by Wiebel and O'Brien, the "indigenous outreach model" to

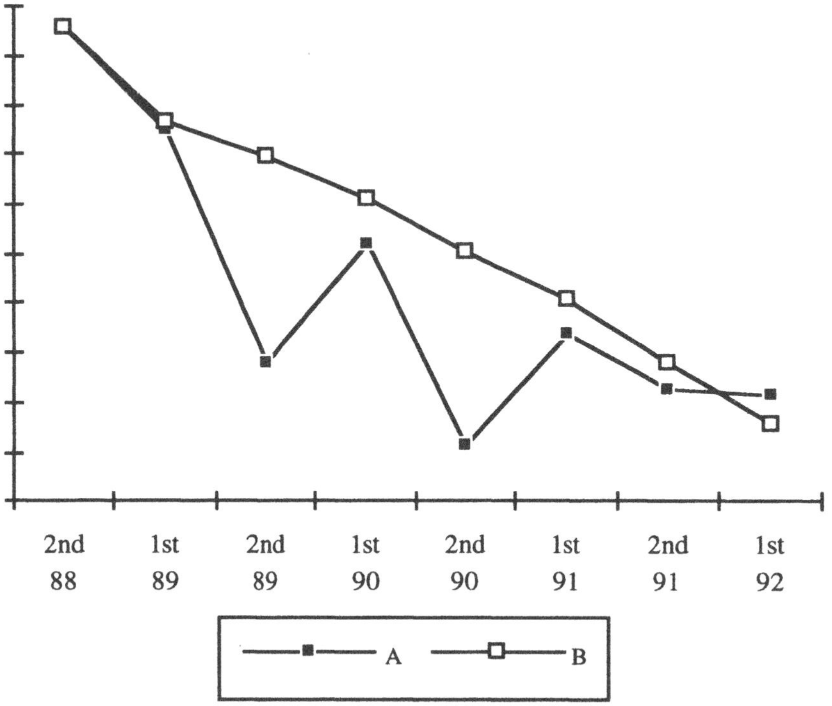


**Fig. 1.** Seroconversion among IDUs. Reduction of High-Risk Injection activity of C.O.I.P. intervention (1982-1992; n=641). A= % injection risk; B= % new HIV-positive
Source: Wiebel and O'Brien, 1993

emphasize the necessary element in the technology for obtaining credibility, trustworthiness, capturing attention and motivating change on the street.

Figure 1 provides a dramatic portrayal of the Chicago results. In the panel of 850 active injectors first measured in 1988 and followed-up at 78.5 % rate until 1992, HIV prevalence at baseline was two to three times the rate found in Chicago treatment settings. Impressive behavior change was demonstrated; since 1988 unsterile injecting dropped in the panel from 95% to a remarkable 15%. The panel was also assayed by blood testing each six months. This provided a "golden standard" rare in prevention studies. The standard confirms the risk reduction behavioral evidence: the annual rate of seroconversion dropped from almost 5% in the first year to less than 2% in 1992. Among these in the panel who still engaged in "dirty sharing", the average six month seroconversion rate was 4.6 % ; among those not engaged in "dirty sharing", the seroconversion rate was 0.4%.

## Further Lowering the Threshold: The Added Value of Self-Organizations

Subcultural mobilization, network interventions and indigenous community field-workers are all distinctive elements that compose this new outreach technology. This technology functions to lower the threshold between active IDUs and helping services. This, in itself, leads to risk reduction. Outreach technology together with methadone maintenance treatment, an older technology for lowering the threshold between addicts and services, provide a powerful system for combating the spread of AIDS in the IDU population. Both technologies involve an active attempt to rehabilitate IDUs before they are completely detoxified. This rehabilitation involves the (self) reorganization of IDU daily life in a way to protect against harmful health consequences (7). By extending the occasions where AIDS is discussed in the context of methadone programming and changing the existing network IDU structure through the intervention of indigenous workers, the daily life of IDUs is changed by increasing the time spent in AIDS awareness activities.

To illustrate, data drawn from the Boule de Neige project (BN) included in Ernst Buning's WHO report on outreach. Starting in Brussels in 1988, the BN methodology has been adopted in the Euregio composing of four distinct, but regionally related cultural areas - French Belgium (Liege); Flemish Belgium (Hasselt), Germany (Aachen) and Netherlands (South Limburg urban area of Maastricht , Heerlen and Sittard). The project has been funded by DG V of the European Commission and coordinated by the Consultation Bureau of Alcohol and Drugs Limburg. The International Institute of

**Table 1.** Background of Drug Users, Risk Behavior of Drug Users and Drug Related Services: Key Variables in Euro-Region Cities

| | Aachen n=40 | Hasselt n=24 | Liège n=119 | S. Limberg n=135 |
|---|---|---|---|---|
| **Background** | | | | |
| Still using drugs (%) | 83 | 91 | 90 | 92 |
| Perceived no HIV infection (%) | 88 | 67 | 86 | 86 |
| Never traded sex for drug (%) | 78 | 80 | 71 | 80 |
| **Risk behaviour** | | | | |
| Increase in shooting style in last 4 weeks (%) | 25 | 8 | 7 | 16 |
| Changed behaviour because of AIDS risk (%) | 65 | 8 | 52 | 66 |
| Often use own syringe (%) | 35 | 17 | 20 | 54 |
| Used condoms last 6 months (%) | 33 | 8 | 37 | 40 |
| **Services** | | | | |
| Condoms easily available (%) | 90 | 67 | 88 | 96 |
| Syringes unavailable (%) | 58 | 33 | 66 | 80 |
| Contact with mental health service (%) | 18 | 13 | 30 | 67 |

Psychosocial and Socioecological Research, University of Limburg, has been employed to provide a basic evaluation.

The BN intervention methodology relies on face-to-face contact. It assess itself in relationship to other more conventional prevention forms of communication such as brochures and television. This methodology was adopted as a common strategy for the Euregio when needle exchange failed to be accepted because of policy differences between the Member States. "Animators" (i.e. key informants of each city) were employed to recruit "jobists". The jobists were required to be indignant, have knowledge about the local drug scene and be willing to act as a role model for the target groups. Fieldworker training was given to the jobist involving drug counseling, sexual counseling, AIDS and research.

Three hundred and eighteen questionnaires were administered in 1992. The baseline assessment revealed noticeable variations between the research sites. This finding supports the view that factors related to the spread of AIDS vary considerably across ecologically distinct units located a small distance apart. Table 1 summarizes some of the important results of the baseline assessment (8). Thus, a key finding related to lowering the threshold is that contact with the mental health services is far greater in South Limburg than the other three sites. Sixty-seven percent of those interviewed in South Limburg had contact with services. This was more than two times as high as found in Liege, the second highest percentage of contact with mental health services (30%). This suggests that a number of aspects characteristic of the Dutch service provision system (i.e. the availability of methadone treatment on demand and needle exchange) increase the magnitude of contact with the targeted population-at-risk. This relatively high level of contact, in turn, seems associated with positive results in risk behavior; on three of four risk behavior variables South Limburg presents the most favorable levels of outcome. Supporting this inference is the case of Hasselt. Here the city with the lowest contact also presents the least favorable outcomes in three of four risk variables.

The case of Aachen illustrates an important emergent property of outreach technology; the stimulation of IDU self-organization. Thus in Aachen, there appears to be a relatively favorable background variables for the prevention of AIDS. However, in contrast to South Limburg, there is a low contact with services. This suggests that background variables as well as services variables are determining the variation in risk behavior observed across the different cities. One plausible explanation is the existence of an unmeasured variable, the level of self-organization of IDUs at baseline, that is interacting with the BN intervention to produce a relatively favorable outcome. The process evaluation report of the BN written by the project's coordinator, A. Penners, documents the movement of German IDUs to Heerlen to utilize Dutch needle exchange services unavailable in their own country (9). Through their own self-organization German IDUs have been able to produce and distribute through their networks critical information pertaining to AIDS prevention. In the Aachen situation, the highest level of awareness of no AIDS infection (88%), an important background variable, is observed despite the low contact of Aachen IDUs with mental health services. This awareness of Aachen IDUs provided in the context of their own self-organizations, stimulates their motivation to seek services that will preserve favorable conditions even if it means traveling across national borders.

A second critical background variable which is most favorably expressed in Aachen is the lowest percentage of still using drugs (83%). This favorable condition may be do to an interaction of a relatively high degree of self-organization with a relatively repressive drug policy. Both more self-control associated with self-organization and policy measures that restrict the access to drugs seem to provide a compensatory mechanism for the lack of contact with services (10). However, this positive interaction effect does not seem to be as strong in the long run as optimal public mental health service provision. This interpretation is evidenced by the uneven distribution of favorable outcomes observable in Aachen as compared to South Limburg.

Further support for the limitations of self-organization and repressive drug policy can also seen in the data. Negative behavior may not be suppressed in a stable way. Positive behaviors such as not sharing syringes do not automatically lead to a decrease in the shooting styles. Self-organization may have their own specific negative effects. Shooting has a special place in German IDU subcultures. German IDUs often call themselves "fixers" highlighting the importance of shooting styles in the subculture. Unlike across the border in The Netherlands where "chasing the dragon" or "chinesing" (i.e. smoking) is the most prevalent using style, the German IDU subculture and its extended self-organizations may counter individual motivations to change from an injecting to a smoking using style (11). This negative outcome is reinforced by repressive drug policy which tends to raise the price and lower the quality of drugs providing an incentive to use more efficient and economical styles of use; i.e injecting instead of smoking.

## Conclusion

Self-organization as a scientific concept has its origin in the study of self-regulating ecosystems (10). A self-organization is a basic ecological unit that produces stores and copies information. As in the case of the Aachen, self-organizations can be seen to complement services and transmit critical information such as where to obtain syringes when services do not distribute them and personal awareness of seropositivity. Self-organizations are structured within the personal networks of IDUs which create the interactional context of the local drug scene. These relatively simple structures have the ability shared with other bioecological simple systems of exhibiting emergent properties such as self-replication which greatly extend their impact on the environment (11). The simple and natural system of IDU self-organizations should not therefore be taken lightly. Profound consequences can and do emerge from such humble structural origins. Thus, recognition of the self-organization component in outreach technology focuses attention on the critical role these simple systems can have in increasing the impact of the technology. The BN project as well as parallel interventions in New York and Chicago are showing that community work that stimulating self-organization through public meetings is an effective intervention in AIDS risk reduction. These community organization activities add value to the existing outreach technology through increasing the scope of their function to include: fighting the stigma attached to IDUs by practically breaking taboos that surround addicts; slowly, and in increments, extending the amount of responsibility that IDUs have for each other; deepening the awareness of AIDS in existing networks and the provision of AIDS care.

The BN project has now concluded its second year of operation. The impact of the project will be able to be better evaluated after the analysis of second measurement of 1993. This second sample is more than two times larger than the 1992 cohort. The outreach teams have been active in the region not only on the streets, but at large rock festivals talking and information sharing about AIDS and drugs. On the basis of the impact evaluation comparing 1992 with 1993 more specific targeted interventions can be done. This highlights an important function of outreach highlighted in the WHO report: intelligence about new developments and needs for effective prevention planning will be possible. The normalization of the situation in the Euregio requires that weak links in the regional prevention system can be detected and resources can be reallocated and mobilized to provide IDUs with the information and resources to effect risk reduction. Furthermore, the BN similar to other outreach projects has stimulated a new epidemiological strategy. Together with methadone services, outreach technology provides a powerful tool for rehabilitating IDUs for risk reduction. Outreach is effective because it works specifically through the mechanism of subcultural change within the network structure of IDUs introducing new norms and discourses. Self-organizations further lower the threshold of outreach technology itself and add new values and possibilities to the system of methadone and outreach services for IDUs.

## References


1. World Health Organization . Global program on AIDS (progress report November 4). In: AIDS, drugs and prostitution, edited by Plant M. Tavistock, London, 1990,
2. Buning E. Outreach work with drug users: an overview. Int J Drug Policy 4:78-80, 1993.
3. Wodak A. A Dutch smorgasbord: research on HIV/AIDS and injecting drug users. Int J Drug Policy 4:5-27, 1993.
4. Houk van de J A R, Coutinho R A, Haastrecht van H J A, et al. . Prevalence and risk factors among drug users and drug using prostitutes in Amsterdam. AIDS 2:55-60, 1988.
5. Friedman S. Going beyond education to mobilizing subcultural change. Int J Drug Policy 4:91-95, 1993.
6. Kaplan C D, Vries de M W, Grund J P C, et al. . Protective factors: Dutch interventions, health determinants and the reorganization of addict life. In: Drug misuse and dependence, edited by Ghodse H A, Kaplan C D, Mann R D. Carnforth, Lancs., 1990, pp 165-176.
7. Kaplan C D, Mercera B, Meulders W A J, et al. . The "Boule De Neige" project: lowering the threshold for AIDS prevention among injecting drug users. Int J Drug Policy 3:170-175, 1992.
8. Grund J P C. Drug use as a social ritual: functionality, symbolism and determinants of self-regulation, IVO Series, Rotterdam, 1993.
9. Grund J P C, Blanken P. From chasing the dragon to chinezen: the diffusion of heroin smoking in The Netherlands, IVO Series, Rotterdam, 1993.
10. Odum H T. Self-organization, transformity and information. Science 242:1132-1139, 1988.
11. Raggia J A, Armentrout S L, Chou H H, et al. . Simple systems that exhibit self-directed replication. Science 259:1282-1287, 1993.

# 19

# Pharmacological Treatment of Addiction: Normalization of Physiology and AIDS Risk Reduction

**M. J. Kreek**
The Rockefeller University, New York, NY, U.S.A.


## Summary

In the United States, it is estimated that over 2 million persons have used heroin at some time and between 500.000 and 1 million are heroin addicts. It is estimated that over 23 million have used cocaine at some time and, of those, between 600.000 and 900.000 use cocaine regularly and are cocaine addicts. Parenteral drug abuse has become the leading risk for development of HIV infection in many regions. Illicit use of other drugs may also contribute to increase risk for HIV infection. The goals of treatment for any addiction must include prevention of abstinence or withdrawal symptoms, reduction or elimination of "drug hunger" or craving and relapse, and normalization of physiological functions which have been disrupted by abuse of the drug. Long-term studies of the physiological effects, in addiction to studies of the safety and efficacy of the long-acting opioid methadone as used in the maintenance treatment of addiction, have shown that this drug successfully addresses each of these three goals. It is assumed that future studies will show that another long-acting opioid, L-alpha-acethyl methadol (LAAM), will have similar desirable effects.

It has been shown that effective methadone maintenance treatment leads to a highly significant reduction in HIV-1 infection as long as patients remain in treatment. In our early study carried out in 1983-1984, we found that less than 10% of patients in effective methadone maintenance treatment has been infected with the HIV-1 virus at a time when over 50-60% of untreated addicts were infected with this AIDS virus. These studies now have been replicated in many parts of the world. Also, our studies have shown that chronic steady dose treatment with methadone allows normalization of many aspects of physiological function which may be profoundly disrupted during cycles of heroin addiction. This normalization includes restoration of function of the neuroendocrine axes which control stress response (the hypothalamic-pituitary-adrenal axis) as well as the reproductive biological responses (the hypothalamic-pituitary-gonadal axis). In addiction, our studies have shown that indices of immune function are disrupted cycles of heroin addiction become normalized during chronic steady dose methadone maintenance treatment, in part probably due to normalization of neuroendocrine function. On-going studies in our Laboratory have shown that the endogenous opioid system may play a physiological role in the modulation of specific indices of immune function such as natural killer cell activity. Therefore, it is essential to restore homeostasis of this endogenous opioid system which is disrupted during cycles of heroin and cocaine addiction. Other diseases now confound the problems of drug addiction and AIDS; they include Hepatitis B, C, and delta, and also tuberculosis. Compromised immune function may alter both the diagnosis and treatment of these disorders. Some drug interactions may occur between methadone as used in maintenance treatment and medications used to treat these other disorders, such as with the anti-tuberculosis agent rifampin. However, by knowledge of such interactions, it is possible to adjust the methadone maintenance treatment dose appropriately to allow appropriate maintenance treatment.

The problems of drug abuse in the United States and worldwide continue to be one of the most important of all medical problems (1,5). The recent 1992 United States National Household Survey on Drug Abuse reported that 54 million Americans are dependent upon tobacco smoking, at least 10 million are heavy ethanol drinkers, defined as consuming five drinks or more in a single occasion on five days or more during the last thirty days (1,2). These two licit dependencies comprise the major chemical dependency problem in the United States at this time. However, use of illicit drugs are also major problems. Because of the frequent use of unsterile needles for self-administration of these agents, the drug use related morbidity and mortality due to other diseases including hepatitis B, C, delta, and especially AIDS, are also enormous (6,11). Although some individuals may use these drugs by the paranasal route or by free-base smoking, the majority still use these drugs by the intravenous or other parenteral routes.

It is estimated that 23.9 million persons have used cocaine at some time, over 3.4 million report occasional use, and 1.3 million current use, with 640,000 reporting frequent or regular cocaine use. Also in this 1992 U.S. National Household Survey, the number reporting having ever used heroin is higher than before, with over 2.7 million reported to have used heroin at some time (1,2). Other studies suggest that between 500,000 and 1 million are currently active heroin addicts. In the recent report of the NIDA National Survey, "Monitoring the Future," published by the University of Michigan group headed by Dr. Lloyd Johnston, it has been reported that 48% of college students and 60% of a large group of young adults are using illicit drugs, with a study population number of over 1,400 and 6,800 respectively (3,4). The primary illicit drug use is marijuana, followed by hallucinogens. In these groups, it was reported that cocaine had been used by 7.9% of college students, and 19.5% of the young adult group (3,4). Also, it was reported that 0.8% of college students and 0.9% of young adults have used heroin. Of considerable importance, 7.3% of college students and 8.9% of young adults reported illicit use of other opiate drugs (3,4). This magnitude of drug abuse in the United States is enormous.

The second factor which has led to increased concern about more appropriate and effective management of addictive diseases, especially heroin addiction and cocaine addiction, has been the advent of the AIDS epidemic. By unbanking and decoding blood specimens collected as part of my research from the late 1960's onward, we were able to determine that HIV infection entered the parenteral drug abusing population in New York City around 1978 (6,11). Subsequent identification of a limited number of babies born to opiate dependent mothers identified somewhat earlier suggested that the HIV-1 infection may have entered the parenteral drug abusing population in 1975 (10). For several years in New York City, intravenous drug abuse has been the major risk factor for development of HIV-1 infection and AIDS. By 1991, over 45% of all new cases of

AIDS were in intravenous drug users (10,11). Thus, the need to effectively prevent and also treat intravenous drug use, primarily heroin addiction and cocaine dependency, is imperative.

For many years, my laboratory has been studying the vulnerability leading to the development of addictive diseases and the neurobiological basis of addiction. We have hypothesized that genetic factors, probably involving multiple genes, may play a role, along with the specific and direct effects of drugs of abuse on physiology, which may lead to persistent or permanent changes contributing to drug-seeking behavior. Because many studies using a variety of drug-free approaches as well as follow-up studies following incarceration have shown that over 70-90% of former long term "hard core" heroin addicts, defined as regular daily use of multiple doses of opiate for one year or more with the development of tolerance, physical dependence, and drug-seeking behavior, will relapse within one year following release from residential drug-free treatment or prison. It became apparent that there were needs for pharmacotherapy for opiate addiction. Similarly, at this time, there still remains a need for development of pharmacotherapy for cocaine dependency and alcoholism.

The goals for pharmacotherapy for an addiction include: 1) the prevention of withdrawal symptoms; 2) the reduction of drug craving; and 3) the normalization of any physiological functions which have been disrupted by drug use. Also, the rationale for use of any drug should include targeting of the agent to a specific site or receptor, or to some physiological or neurochemical process known to be altered by use of the addictive agent. All of these criteria have been met by methadone as used in the maintenance treatment of heroin addicts. There are now several pharmacological agents used for the treatment of opiate dependency (12,15). These include the most widely used and most effective agent, racemic dl-methadone, and also to a very limited extent, its active enantiomer, l-methadone (12,15). In addition, another longer acting congener of methadone has recently been approved by the United States Food and Drug Administration following studies conducted under the auspices of the National Institute on Drug Abuse, l-alpha-acetylmethadol (LAAM). All of these agents are specific opioid agonists. In addition, the specific opioid antagonist, naltrexone, has been approved for use since 1984, but has been shown to be generally ineffective in an unselected group of heroin addicts, with less than 15-20% remaining voluntarily in treatment for one year, and thus benefiting from such management. However, it may be more effective for shorter term treatment of specific patient subgroups. Also, at this time, studies are being conducted in the United States to determine the possible utility and indications for use of the partial opioid agonist, buprenorphine, sometimes referred to as a mixed opioid agonist-antagonist which acts, however, primarily as a $\mu$ opioid agonist.

In early 1964, under the leadership of Professor Vincent P. Dole, the first studies on the possible use of methadone for the maintenance treatment of addiction were initiated at the Hospital of the Rockefeller Institute for Biomedical Research (now called the Rockefeller University) (12,15). At that time, there were no sensitive and specific techniques for measuring plasma and urine levels of either licit or illicit opiate drugs; rather, careful clinical observations had to be used to determine both the pharmacokinetic and pharmacodynamic properties of each agent. There was some limited experience of use of methadone in the treatment of chronic pain, which suggested that methadone might have long-acting properties, since accumulation seemed to occur when multiple doses were given each day. Also, methadone had been used in short-term detoxification on a limited basis at the United States Public Health Service resource in Lexington, Kentucky, as well as at the Bellevue Hospital and Metropolitan Hospitals in New York City.

In considering the alternatives for pharmacotherapy at that time, our research group thought that methadone was most likely to be efficacious because it was orally effective and seemed to have long-acting properties, two features essential for any medication to be used in the chemotherapy of addiction. Subsequent studies in addictive disease populations proved those assumptions to be correct: Methadone was found to be orally effective; to have an onset of action within 30 minutes; and to have a duration of action of 24-36 hours, in sharp contrast to heroin, which was fully effective only when given by the intravenous route; to have an immediate onset of action, and a duration of action of only 3-4 hours. Of even more importance, when the dose of methadone was carefully selected to be within the degree of tolerance developed by that individual, no euphoric signs or symptoms, and in fact, no narcotic effects, were ever observed or experienced subjectively by the patient. This is in sharp contrast to administration of heroin, which is given in doses sufficient to prevent withdrawal symptoms for 3-4 hours, and results in euphoria for 1-3 hours. Also of great importance, methadone was found to have a duration of action of about 24 hours with prevention of withdrawal symptoms for that time, whereas the duration of heroin would be only 3-4 hours (12,15).

By 1972, my laboratory had developed sensitive and specific techniques for measuring plasma levels of methadone using gas liquid chromatographic methods (16,17). Using these techniques, we were able to determine that the racemic methadone had a half-life of around 24 hours in parallel with the pharmacodynamics in preventing both abstinence symptoms and drug hunger (16). However, altered pharmacokinetics were found when agents such as rifampin, phenytoin, or ethanol, which alter drug metabolism, were used, or when the patient was pregnant, in which setting methadone metabolism is also altered (17,20). In subsequent studies in which we used stable isotope-labeled methadone, with each enantiomer labeled with different amounts of deuterium so that each enantiomer could be followed simultaneously after intravenous injection or oral administration of a single daily dose in the stable isotopically labeled form, the techniques of gas chromatography-chemical ionization mass spectrometry with selective ion monitoring were used. We found that the plasma beta terminal half-life of the active l-(s) enantiomer of methadone was significantly longer than that of the inactive d+(l) enantiomer (21,25).

To date, over 115,000 persons have been successfully treated with methadone maintenance treatment in good programs in which adequate doses were used, and access to counseling and other rehabilitation, as well as medical services, are available (13,15). Many studies have shown that there is a 70-80% voluntary retention for one year or more (13,15). Also, when adequate doses are used, the continuing use of illicit heroin after six months in treatment falls to below 15% (13,15). The action of methadone is that it prevents drug hunger, it blocks the euphoric effects of short-acting narcotics, and of great importance, it has also been shown that it allows normalization of physiological interventions which have been disrupted by chronic use of the short-acting narcotics. The mechanism of action is that this long-acting opioid agonist provides steady state perfusion of specific opioid receptors in over a 24 hour dosing interval. Of great successful methadone maintenance program before the advent of the AIDS epidemic in 1978, most were HIV-negative in 1983-84, with only 9% showing HIV positivity at that time, as contrasted to street heroin addicts and those heroin addicts entering methadone maintenance treatment in 1984, with 50-60% of that group being already HIV-infected (6,10). Thus, our studies and many other subsequent studies have given strong support to the fact that effective methadone maintenance treatment highly significantly protects former heroin addicts from infection with the HIV virus by virtue of essentially stopping all use of illicit opiates and thus the major reason for exposure to the HIV-1 virus by use of shared unsterile needles.

The natural history of a variety of diseases may be affected by heroin addiction, as well as by cocaine addiction and alcoholism. Immunological function has been shown to be markedly deranged in the setting of active heroin addiction, and may be perturbed, in part due to the chronic use of drugs injected by unsterile needles, with exposure to a wide variety of foreign substances and infectious agents, in part due to general life-style, including poor nutrition and generally poor health status (26,28). However, there are increasing evidences, both from laboratory as well as clinical research, that opiates, as well as other drugs of abuse, including cocaine and alcohol, may significantly alter immune function. In addition to AIDS, one of the most common sets of diseases in active parenteral drug abusers is hepatitis B, hepatitis C, and hepatitis delta (26,33). From our earliest work in the 1960's up until around 1985, our group and many others working in the United States found that over 90% of all intravenous drug users have been infected with the hepatitis B virus. Of great interest, we have found that in most cases, the hepatitis B markers observed are antibodies reflecting infection, antibodies to hepatitis B virus core reflecting prior infection, that is, anti-HBc, or alternatively, protective antibodies also reflecting prior infection, anti-HBs. However, 5-15% of all heroin addicts have been found to have hepatitis B antigenemia, that is, on-going infection with virus. We have found that during long-term methadone treatment, the prevalence of hepatitis B antigenemia falls to zero, suggesting that improved immune function may allow clearance of the virus to occur. From 1985 onward, we have begun to note a decreasing prevalence of hepatitis B markers in active intravenous drug users, suggesting that AIDS risk reduction education has effectively reduced the spread of hepatitis B along with, hopefully, the spread of HIV infection.

Hepatitis delta is caused by a circular RNA viroid-like particle which depends upon hepatitis B viral replication for coating material which allows ready entry into hepatocytes, and thus leads to infection. In various studies, we have found that delta hepatitis has infected from 5% to over 30% of intravenous heroin addicts (10,30,32). In recent studies, we have shown that the most infectious form of this viroid-like RNA agent, delta hepatitis, exists in the presence of delta agent antigenemia, indicating active replication is more common in intravenous drug users with AIDS than in those without AIDS (10). Similarly, hepatitis B antigenemia is more common in intravenous drug users with AIDS, as compared with those without AIDS (10). These findings suggest that the profound immune compromise caused by the HIV-1 virus alters the natural history of both hepatitis delta and hepatitis B, findings of extreme importance in the opiate addict.

Also, early studies from our laboratory have shown that concomitant use of cocaine along with heroin predisposed more intravenous drug users for the development of HIV-1 infection early in the epidemic (34). Presumably, the complicating behaviors of intravenous heroin use as well as concomitant cocaine dependence increases the predisposition to HIV-1 infection exposure. However, the possibility of additional alterations to immune function by cocaine use, which may have contributed to HIV infection, must also be considered (34,35).

Several studies, including our own, have shown that natural killer cell activity may be significantly reduced in intravenous heroin addicts. Also, studies, including our own, have shown that total lymphocyte counts may be increased in heroin addicts (26,28,36). Absolute numbers of both CD-4 and CD-8 cells have been shown to be increased in heroin addicts without HIV-1 infection. Since several laboratory studies have suggested that opiates and possibly the endogenous opioids may significantly reduce natural killer cell activity, as well as alter other indices of immune function, our laboratory conducted studies in well-rehabilitated methadone maintained patients, all subjects who had been in continuous treatment for over eleven years, with moderate to high doses of methadone (36). If, in fact, methadone, which is a very long-acting opioid in man, caused immune disruption, one would expect to see the most profound immune disruption in this population of former heroin addicts who had been in continuous treatment with methadone for a long period of time, since both the total amounts of opiate received and the half-life would have yielded much greater exposure of the immune system to opioids. In fact, what was found was that specific indices of immune function had become normal in these long-term methadone maintained patients. Normal absolute levels of CD-4 and CD-8 cells and normal levels of immunoglobulins were found (36). Of greatest clinical importance, since natural killer cells play a role in many viral infections such as hepatitis B, and may play a role in some of the neoplastic disorders complicating HIV infection, natural killer cell activity was found to be normal in a long-term methadone maintained patient (36). In vitro studies in our laboratory had also shown that methadone and also naloxone have no effects on natural killer cell activity until extraordinarily high concentrations of drug are added, much greater than patients will ever be exposed in vivo (37,38). Other evidence from our laboratory has suggested that this normalization of immune function during long-term methadone maintenance of heroin addicts is possibly directly due to normalization of neuroendocrine function which occurs in that setting (13,27,28,36).

Very recent work from our laboratory has suggested that chronic administration of cocaine may also alter the important hypothalamic-pituitary-adrenal axis in humans (39). This has led our laboratory to model the "binge pattern" of cocaine administration in rodents, and to determine whether or not binge pattern cocaine administration will affect the endogenous opioid system. Several recent exciting findings have been made by our laboratory, suggesting that indeed the endogenous opioid system is profoundly perturbed during exposure to chronic "binge pattern" cocaine administration, including increased density or up-regulation of both $\mu$ and k opioid receptors in those regions of the brain where dopaminergic terminals are the most abundant, and areas which have been implicated in drug self-administration and locomotor activity, including the nucleus accumbens and the caudate putamen region of the striatum (40,41). Also, we have found that messenger RNA for the prodynorphin gene is enhanced in a region of the brain where dopaminergic terminals are abundant, the caudate putamen region of the striatum, along with increased levels of dynorphin peptides in other specific brain regions (42,43). However, transient, but no persistent changes in preproenkephalin mRNA have been found in this setting (43,44). Further studies are on-going in our laboratory to determine whether disregulation of the opioid receptors, as reflected by alterations in levels of mRNA for each of the recently cloned opioid receptor genes, also occurs in this setting.

Therefore, in these studies, we have found that the drugs of abuse, including opiates and also cocaine, may disrupt the endocrine system, and this may in turn cause disruption of the endogenous opioid system, or may include alteration of immune function, may alter the initial or continuing immune response to viral agents, including possibly that of HIV infection. On the other hand, we have found that long-term treatment with methadone for ten years or more is medically very safe, as well as effective (45).

## References

1. National Household Survey on Drug Abuse: Main Findings 1991, Substance Abuse and Mental Health Administration, DHHS Publication No. (SMA) 93-1980, Rockville, MD, 1993.
2. Preliminary Estimates from the 1992 National Household Survey on Drug Abuse, Substance Abuse and Mental Health Administration, Advance Report n.3, June , Rockville, MD, 1993.
3. Johnston L D, O'Malley P M, Bachman J G. National Survey Results on Drug Use from Monitoring the Future Study, 1975-1992, Vol. 1, National Institute on Drug Abuse NIH Publication No. 93-3597, Rockville, MD, 1993.
4. Johnston L D, O'Malley P M, Bachman J G. National Survey Results on Drug Use from Monitoring the Future Study, 1975-1992, Vol. 2, National Institute on Drug Abuse NIH Publication No. 93-3597, Rockville, MD, 1993.
5. Estimates from the Drug Abuse Warning Network: 1992 Estimates of Drug-Related Emergency Room Episodes, Substance Abuse and Mental Health Services Administration, Advance Report N. 4, September, Rockville, MD, 1993.
6. Des Jarlais D C, Marmor M, Cohen H, Yancovitz S, Garber J, Friedman S, Kreek M J, Miescher A, Khuri E, Friedman S M, Rothenberg R, Echenberg D, O'Malley P O, Braff E, Chin J, Burtenol P, Sikes R K. Antibodies to a retrovirus etiologically associated with Acquired Immunodeficiency Syndrome (AIDS) in populations with increased incidences of the syndrome. Morbidity and Mortality Weekly Report 33:377-379, 1984.
7. Novick D M, Khan I, Kreek M J. Acquired immunodeficiency syndrome and infection with hepatitis viruses in individuals abusing drugs by injection. United Nations Bulletin on Narcotics 38:15-25, 1986.
8. Novick D, Kreek M J, Des Jarlais D, Spira T J, Khuri E T, Raghunath J, Kalyanaraman V S, Gelb A M, Miescher A. Antibody to LAV, the putative agent of AIDS, in parenteral drug abusers and methadone-maintained patients: Abstract of clinical research findings: Therapeutic, historical and ethical aspects. In: Problems of Drug Dependence, 1985; Proceedings of the 47th Annual Scientific Meeting of the Committee on Problems of Drug Dependence, NIDA Research Monograph Series, edited by Harris L S. DHHS Publication No. (ADM) 86-1448, Rockville, MD, 1986, pp 318-320.
9. Des Jarlais D C, Friedman S R, Novick D M, Sotheran J L, Thomas P, Yancovitz S R, Mildvan D, Weber J, Kreek M J, Maslansky R, Bartelme S, Spira T, Marmor M. HIV I Infection among intravenous drug users in Manhattan, New York City 1977 to 1987. J Am Med Assoc 261:1008-1012, 1989.
10. Kreek M J, Des Jarlais D C, Trepo C L, Novick D M, Abdul-Quader A, Raghunath J. Contrasting prevalence of delta hepatitis markers in parenteral drug abusers with and without

Contrasting prevalence of delta hepatitis markers in parenteral drug abusers with and without AIDS. J Infect Dis 162:538-541, 1990.
11. Kreek M J. HIV infection and parenteral drug abuse: ethical issues in diagnosis, treatment, research and the maintenance of confidentiality. In: Proceedings of the Third International Congress on Ethics in Medicine - Nobel Conference Series, edited by Allenbeck P, Jansson B. Raven Press, New York, NY, 1990, pp 181-187.
12. Dole V P, Nyswander M E, Kreek M J. Narcotic blockade. Arch Intern Med 118:304-309, 1966.
13. Kreek M J. Rationale for maintenance pharmacotherapy of opiate dependence. Addictive States, Raven Press, New York, 1992.
14. Kreek M J. Epilogue: Medical maintenance treatment for heroin addiction, from a retrospective and prospective viewpoint. In: State Methadone Maintenance Treatment Guidelines, Office for Treatment Improvement, Division for State Assistance , 1992, pp 255-272.
15. Kreek M J. The addict as a patient. In: Substance Abuse: A Comprehensive Textbook, edited by Lowinson J H, Ruiz P, Millman R B, Langrod J G. Williams & Wilkins, Baltimore, ML, 1992, pp 997-1009.
16. Kreek M J. Plasma and urine levels of methadone. N Y State J Med 73:2773-2777, 1973.
17. Kreek M J, Gutjahr C L, Garfield J W, Bowen D V, Field F H. Drug interactions with methadone. Ann N Y Acad Sci 281:350-370, 1976.
18. Kreek M J, Garfield J W, Gutjahr C L, Giusti L M. Rifampin-induced methadone withdrawal. New Engl J Med 294:1104-1106, 1976.
19. Tong T G, Pond S M, Kreek M J, Jaffery N F, Benowitz N L. Phenytoin-induced methadone withdrawal. Ann Intern Med 94:349-351, 1981.
20. Pond S M, Kreek M J, Tong T G, Raghunath J, Benowitz N L. Altered methadone pharmacokinetics in methadone-maintained pregnant women. J Pharmacol Exp Ther 233:1-6, 1985.
21. Hackey D L, Kreek M J, Mattson D H. Quantitative analysis of methadone in biological fluids using deuterium-labeled methadone and GLC-chemical-ionization mass spectrometry. J Pharm Sci 66:1579-1582, 1977.
22. Kreek M J, Hackey D L, Klein P D. Stereoselective disposition of methadone in man. Life Sci 24:925-932, 1979.
23. Nakamura K, Hackey D L, Kreek M J, Irving C S, Klein P D. Quantitation of methadone enantiomers in humans using stable isotopo-labeled 2H3, 2H5, 2H8 methadone. J Pharm Sci 71:39-43, 1982.
24. Kreek M J, Bencsath F A, Field F H. Effects of liver disease on urinary excretion of methadone and metabolites in maintenance patients: Quantitation by direct probe chemical ionization mass spectrometry. Biomed Mass Spectrm 7:385-395, 1980.
25. Kreek M J, Bencsath F A, Fanizza A, Field F H. Effects of liver disease on fecal excretion of methadone and its unconjugated metabolites in maintenance patients: Quantitation by direct probe chemical ionization mass spectrometry. Biomed Mass Spectrm 10:544-549, 1983.
26. Kreek M J, Dodes L, Kane S, Knobler J, Martin R. Long-term methadone maintenance therapy: Effects on liver function. Ann Intern Med 77:598-602, 1972.
27. Kreek M J. Medical safety and side effects of methadone in tolerant individuals. J Amer Med Ass 223:665-668, 1973.
28. Kreek M J. Medical complications in methadone patients. Ann N Y Acad Sci 322:110-134, 1978.
29. Novick D M, Gelb A M, Steger R J, Yancovitz S R, Adelsberg B, Chateau F, Kreek M J. Hepatitis B serologic studies in narcotic users with chronic liver diseases. Am J Gastroenter 75:111-115, 1981.
30. Novick D M, Enlow R W, Gelb A M, Stenger R J, Fotino M, Winter J W, Yancovitz S R, Schoenberg M D, Kreek M J. Hepatic cirrhosis in young adults: Association with adolescent

31. Novick D M, Kreek M J, Arns P A, Lau L L, Yancovitz S R, Gelb A M. Effects of severe alcoholic liver disease on the disposition of methadone in maintenance patients. Alcoholism: Clin Exp Res 9:349-354, 1985.
32. Novick D M, Farci P, Karayiannis P, Gelb A M, Stenger R J, Kreek M J, Thomas H C. Hepatitis D virus antibody in HBs Ag-positive and HBs Ag-negative substance abusers with chronic liver disease. J Med Virol 15:351-356, 1985.
33. Novick D M, Farci P, Croxson S T, Taylor M B, Schneebaum C W, Lai E M, Bach N, Senie R T, Gelb A M, Kreek M J. Hepatitis delta virus and human immunodeficiency virus antibodies in parenteral drug abusers who are hepatitis B surface antigen positive. J Infect Dis 158:795-803, 1988.
34. Novick D M, Trigg H L, Des Jarlais D C, Friedman S R, Vlahov D, Kreek M J. Cocaine injection and ethnicity in parenteral drug users during the early years of the human immunodeficiency virus (HIV) epidemic in New York City. J Med Virol 29:181-185, 1989.
35. Novick D M, Des Jarlais D C, Kreek M J, Spira T J, Friedman S R, Gelb A M, Stenger R J, Schable C A, Kalyanaraman V S. The specificity of antibody tests for human immunodeficiency virus in alcohol and parenteral drug abusers with chronic liver disease. Alcoholism: Clin Exp Res 12:687-690, 1988.
36. Novick D M, Ochshorn M, Ghali V, Croxson T S, Mercer W D, Chiorazzi N, Kreek M J. Natural killer cell activity and lymphocyte subsets in parenteral heroin abusers and long-term methadone maintenance patients. J Pharmacol Exp Ther 250:606-610, 1989.
37. Ochshorn M, Novick D M, Kreek M J. In vitro studies of methadone effect on natural killer (NK) cell activity. Israel J Med Sci 26:421-425, 1990.
38. Ochshorn-Adelson M O, Novick M D, Khuri E, Albeck H, Hahn E F, Kreek M J. Effects of the opioid antagonist naloxone on human natural killer cell activity in vitro. Israel J Med Sci in press: 1993.
39. Kreek M J. Multiple drug abuse patterns and medical consequences. In: Psychopharmacology: The Third Generation of Progress, edited by Meltzer H Y. Raven Press, New York, 1987, pp 1597-1604.
40. Unterwald E M, Horne-King J, Kreek M J. Chronic cocaine alters brain mu opioid receptors. Brain Res 584:314-318, 1992.
41. Unterwald E M, Cox B M, Kreek M J, Cote T E, Izenwasser S. Chronic repeated cocaine administration alters basal and opioid-regulated adenyl-cyclase activity. Synapse 15:33-38, 1993.
42. Spangler R, Unterwald E M, Branch A, Ho A, Kreek M J. Chronic cocaine administration increases prodynorphin mRNA levels in the caudate putamen of rats. In: Problems of Drug Dependence, 1992; Proceedings of the 54th Annual Scientific Meeting of the College on Problems of Drug Dependence. NIDA Research Monograph Series, edited by Harris L S. DHHS Publication No. (ADM) 93-3505, Rockville, MD, 1993, pp 142.
43. Spangler R, Unterwald E M, Kreek M J. 'Binge' cocaine administration induces a sustained increase of prodynorphin mRNA in rat caudate-putamen. Mol Brain Res 19:323-327, 1993.
44. Branch A D, Unterwald E M, Lee S E, Kreek M J. Quantitation of preproenkephalin mRNA levels in brain regions from male Fischer rats following chronic cocaine treatment using a recently developed solution hybridization procedure. Mol Brain Res 14:231-238, 1992.
45. Novick D M, Richman B L, Friedman J M, Friedman J E, Fried C, Wilson J P, Townley A, Kreek M J. The medical status of methadone maintained patients in treatment for 11-18 years. Drug Alc Dep 33:235-245, 1993.

# 20

# Methadone Treatment and the Epidemiology of AIDS in the European Community

**M. Reisinger**
Commission of the European Community, European Monitoring Centre for Drugs and Drug Addiction, (EMCDDA), Brussels, Belgium

## Summary

All the data currently available seem to indicate that the fact that methadone treatment was limited may one day be seen as an inexcusable error of judgement (much more serious than the one committed in some countries with regard to hemophiliacs and transfusion recipients) which will cost the lives of hundreds of thousands of persons and also wreak havoc on the health care budgets of several European countries

## More AIDS Cases Diagnosed among Drug Users than in any Other Group

Since 1991, intravenous drug users (IDUs) constitute the most numerous category of new AIDS cases in Europe. More than 40 percent of the AIDS cases diagnosed in the twelve European Community countries during 1992 were IDUs. At this time, 31,000 AIDS cases had been diagnosed among the IDUs in the twelve EC countries. And this is only the tip of the iceberg. The actual extent of the AIDS epidemic depends on the number of persons who are seropositive. Unlike the number of AIDS cases, this figure cannot be determined at present. However, it can be estimated that the number of HIV positive drug users is between 300,000 and 500,000.

There is another essential fact to be taken into account: it is among IDUs that the increase in the number of AIDS cases is the most significant. For several years now, AIDS has been first and foremost an illness affecting drug users and that this tendency is bound to be accentuated in the years to come.

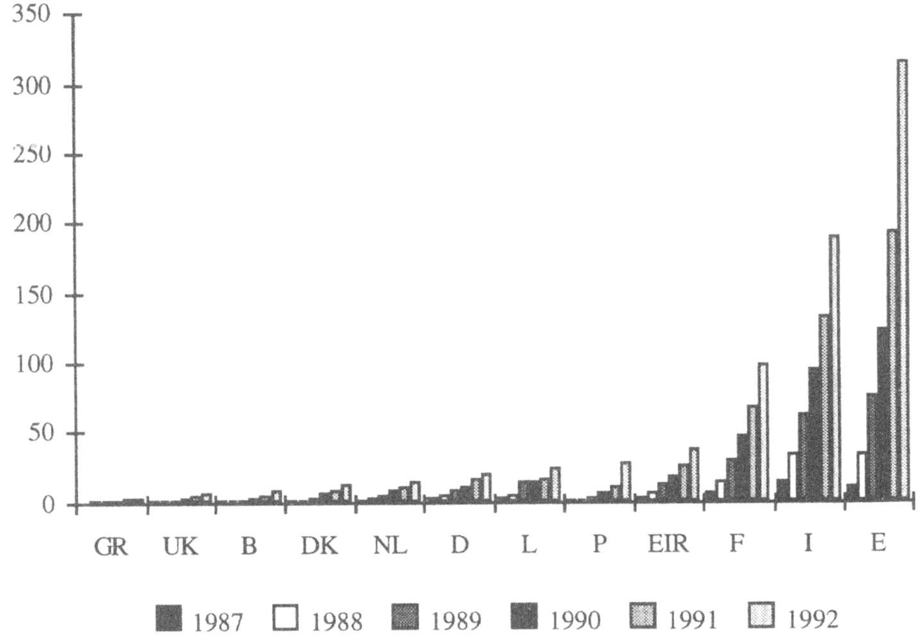


**Fig. 1.** Cumulative AIDS cases in the IDU transmission group (rates per million population)

## Differences among the 12 EC Countries

The extent to which AIDS has spread among drug users varies widely in the twelve European Community countries, as shown in Fig.1. The cumulative rates of HIV-infected intravenous drug users ranges from 3 to 315 per million population, thus in a ratio of 1:100.

In examining this diagram, one cannot but wonder what the causes of these variations in the number of AIDS cases among IDUs in the EC countries might be. The differences in the number of IDUs in each country are not sufficient to account for the differences in the number of IDUs infected with AIDS. Moreover, the date of the emergence of the AIDS epidemic is about the same in all the countries. Thus, nor can this factor alone explain the different evolution of HIV infection among drug users.

This leads one to wonder if these differences might not be due to the manner in which drug users are treated within the various EC countries. Such a hypothesis has been advanced but not yet confirmed, perhaps because it has never been tested in a systematic way (1-3). Research in this field tends to concentrate on IDUs' behavior rather than on the way they are treated. IDUs' reactions towards AIDS prevention campaigns, sterile syringes and treatments seem to be rather similar from one country to another. Nevertheless more than 90 percent of the AIDS cases among drug users are found in three countries: France, Italy and Spain, which account for just 45 percent of the population of the European Community. Is this disproportionate share related to differences in the availability of methadone treatment in the twelve EC countries ? This is the question that will be addressed in this paper.

Comparisons at national level reveal meaningful differences in the percentage of IDUs being treated with methadone in each EC country. In Italy, methadone treatments are for the most part extremely brief detoxification treatments, which cannot have a lasting influence on the behavior of IDUs. That is why we only considered in this country the number of patients who undergo methadone treatment for more than 60 days (approximately 10% of the total number of patients receiving methadone).

An objection raised against efforts to establish a link between methadone treatment and AIDS among IDUs is the fact that the number of diagnosed AIDS cases does not give a true picture of the actual extent of the epidemic, which depends, rather, on the number of seropositives. Unfortunately, statistics on seropositivity are generally insufficient. This makes a valid comparison among countries impossible.

AIDS cases, on the other hand, are recorded much more systematically and thus provide a valid assessment of the extent of HIV infection with a lapse of 7 to 10 years. Consequently, the prevalence and incidence of AIDS cases among IDUs in 1990, 1991 and 1992 illustrate the spread of the virus at the beginning of the period 1983-1990, which is the crucial period for the AIDS epidemic among IDUs in Europe. Definitive conclusions concerning the AIDS epidemic among IDUs during this period, as well as how it might be related to methadone treatment, cannot be drawn until 1997.

It is, nevertheless, possible to analyze the epidemiological data and to formulate hypotheses as to a possible link between the spread of AIDS and methadone treatment. Current statistics on the prevalence and the incidence of AIDS among IDUs in the European Community make it possible to distinguish three groups of countries (see Table 1).

GROUP I: Countries with a high prevalence of AIDS among IDUs (more than ±100 cases per million inhabitants): Spain, France, Italy.

GROUP II: Countries with a low prevalence (less than 50 cases per million inhabitants) and an increasing incidence (more than a 50 percent increase in cases between 1990 and 1992): Belgium, Ireland and Portugal.

GROUP III: Countries with low prevalence and stable incidence (a less than 30 percent increase in cases between 1990 and 1992): Denmark, Netherlands, United Kingdom. Greece and Germany also belong to this group. We will see later why they do not appear in the Table.

In 9 out of 11 countries (Luxemburg was not taken into account due to its very small population) the differences in the prevalence and incidence of AIDS among IDUs can be related to the percentage of IDUs undergoing methadone maintenance treatment. During the 1980's, the countries in GROUPS I and II - with a high prevalence or increasing incidence - had a methadone maintenance treatment capacity sufficient for only 0.05 to 3 percent of the IDUs. The difference between the high prevalence in the GROUP I countries, whatever the case may be, the relative importance of methadone treatment and syringe exchange in preventing AIDS among IDUs is a rather academic question. From the standpoint of public health, if these two measures prove effective, efforts should be made to develop both simultaneously, to the maximum extent, rather than choosing between them as is still often the case.

**Table 1.** AIDS and Methadone Treatment (MT) in 9 E.C. Countries

<table>
<tr><th></th><th colspan="2">AIDS CASES AMONG IDU</th><th>PERCENTAGE OF IDU IN MT</th><th>COUNTRY</th></tr>
<tr><td>GROUP 1</td><td colspan="2">HIGH PREVALENCE<br>>100 cases per million inhabitants</td><td rowspan="2">0.05-3%</td><td>FRANCE<br>ITALY<br>SPAIN</td></tr>
<tr><td>GROUP 2</td><td rowspan="2">LOW PREVALENCE<br><50 cases per million inhabitants</td><td>HIGH INCIDENCE<br>growth rate >50%</td><td>BELGIUM<br>IRELAND<br>PORTUGAL</td></tr>
<tr><td>GROUP 3</td><td>LOW INCIDENCE<br>growth rate < 30%</td><td>15-30%</td><td>DENMARK<br>NETHERLANDS<br>U. KINGDOM</td></tr>
</table>

## The Supply of Methadone Treatment is Inferior to the Demand

In all countries of the European Community, the percentage of IDUs receiving methadone treatment is relatively low: between 0 and 30 percent. Some countries, such as Germany and Spain, have begun to increase the number of methadone treatment slots available. Nonetheless, almost everywhere the supply is inferior to the demand on the part of IDUs. Everywhere methadone is used, there are waiting lists to enter the programs. In several countries, the IDUs have submitted petitions and organized demonstrations - for example in Berlin, Paris and Porto - to demand the expansion of such programs.

The limited supply of methadone treatment slots is due to many factors. The main obstacle is of a legislative nature. Greek law prohibits all treatment except detoxification and that provided by therapeutic communities. Until the referendum of April 1993, Italian law (as well as regional applicative measures) imposed strict conditions for provision of methadone and prohibited doctors in private practice from using such treatments. The Spanish law of 1985 put a stop to methadone treatment, which was just beginning to be used there. Between 1985 and 1986, the number of drug users being treated with methadone decreased from 5000 to 1000 and it was not until five years later that the number of patients reached its initial level.

The attitude of specialists in the field of drug abuse also plays an important role in limiting the availability of methadone treatment. Dr. Costa, from Lisbon, said, for example, that, since 1987, a "drug-free philosophy" has shaped Portugal's system for treating drug users (6), this despite the alarming increase in the HIV seropositive rate among IDUs. The dominant philosophy of French specialists in the field has also prevented methadone from being considered a bona fide therapy, instead of as treatment still in the experimental stages.

In general, the medical practice is not very enthusiastic about treating drug users, who are difficult patients, especially if they have reached a point of advanced social marginalisation. Moreover, both law-makers and medical organizations have taken it upon themselves to impede those who seek to treat drug addicts with maintenance medication. This has occurred in Belgium, Italy, Germany, Ireland and other countries.

In addition to these obstacles, there is also the negative attitude of pharmacists who are sometimes responsible for the administration of methadone prescriptions, but who are often inadequately informed about it. Public hostility to the setting up of drug addiction treatment centers also impedes the use of methadone treatment, as is the case in Dublin, for example.

How is it that the availability of methadone treatment is kept systematically lower than the demand? The most radical means to achieve this end is to prohibit methadone, as is done in Greece. Another method used is to restrict methadone as an experimental treatment. This has been the case for 20 years in France, where the total number of patients being treated with methadone is 52 (it is expected to be raised at 100 patients at the end of 1993). There are strict quotas in several other countries, such as Luxemburg, where only 25 addicts are admitted in methadone treatment. In several Land of Germany methadone is still reserved exclusively for HIV positive addicts, as it was in Edinburgh before 1988, where there were cases of IDUs infecting themselves voluntarily in order to be treated (Mordaunt J., Communication at the Fourth International Conference on the Reduction of Drug Related Harm, Rotterdam 14-18 March,1993). Another means resorted to consists of imposing limits on the duration for the treatment and the doses supplied, as is done on a large scale in Italy. Establishing rigid rules and regulations for methadone programs is another method used to limit the demand (and thus the offer) of such treatments. The waiting list, registration formalities, procedures for daily provision of methadone doses, the fact that one must turn to an institution instead of one's own doctor, the requirement to undergo urinalysis under surveillance, the exclusion from treatment in case of relapse - all these measures have the effect of putting off those who would otherwise seek treatment.

## Conclusion

Is this situation to be considered acceptable, or rather, should it be denounced as an inadmissible refusal to provide treatment to persons in need of it? Are we to await definitive proof of methadone's positive role in preventing AIDS among IDUs, before making such treatment sufficiently available to satisfy the demand? It should not be forgotten that science's "truths," contrary to those of religion, are, by their very nature, to be considered temporary; it is therefore senseless to wait for some "ultimate truth" to be revealed before taking action.

Is it not the duty of public health officials to place their bets on what appears to be the most reasonable and most humane course of action, based on existing knowledge?

The question today is: what is riskier, to treat drug users with methadone or not? All the data currently available seem to indicate that the fact that methadone treatment was limited may one day be seen as an inexcusable error of judgement (much more serious than the one committed in some countries with regard to hemophiliacs and transfusion recipients) which will cost the lives of hundreds of thousands of persons and also wreak havoc on the health care budgets of several European countries (7).

## References


1. Buning E, van Brussel G. Combatting AIDS among intravenous drug users in Amsterdam, Presentation for EC-ad hoc Group AIDS, Brussels, 1988.
2. Reisinger M. Influence de la politique de soins des toxicomanes sur l'épidémiologie du SIDA. Psychotropes VI,1:51-59, 1990.
3. AA.VV. . AIDS and drug addiction in the European Community, European Monitoring Centre for Drugs and Drug addiction Preparatory work program, Brussels, 1993.
4. Malliori M, et al. . HIV-1 infection in intravenous drug users in Greece. In: AIDS and drug addiction in the European Community, edited by Reisinger M. European Monitoring Centre for Drugs and Drug addiction Preparatory work program, Brussels, 1993, pp 105-113.
5. AA.VV. . Abstract Book IXth International Conference on AIDS, Berlin, 1993.
6. Costa A. Treatment of heroin addicts and HIV epidemiology in Portugal. In: AIDS and drug addiction in the European Community, edited by Reisinger M. European Monitoring Centre for Drugs and Drug addiction Preparatory work program, Brussels, 1993, pp 175-180.
7. Lambert D C. Le cout mondial du SIDA, CNRS Editions, Paris, 1992.

# 21
# Public Health in Face of Drug Addiction and AIDS

**C. Vetere**
UNICRI, Rome, Italy

## Summary

Public Health should represent the merging point between scientific knowledge of Medicine and Social Sciences and the National and Local Governments. The Chief Medical Officer should act as an health educator for Ministers and Politicians, because in face of health emergencies a negative attitude by part of decision-making people can provoke useless and dangerous measures. In the same time Public Health is deeply involved into Health Care problems and environmental hygiene which means a lot of administrative responsibilities and the risk of over-dependence from ritualistic measures often adopted in order to avoid criticism and the intervention of the Court.
Drug addiction with its moral and political connotations represents an important example of the prevalence of ethical or personal judgements on objective evaluation of the results of scientific research. Methadone can be signaled as the proof of the consequences of such a situation: even when it is allowed restrictions are so many and attitude so punitive that it would be like interrupt insulin to a diabetic patient who eats too many sweets! Even the prevention of overdose through the larger distribution of Naloxone vials is not actively encouraged by the same authorities who correctly suggest to insulin treated diabetic to bring some sugar lumps. Aids provoked the coming out of all contradictions, discriminations, fears, quarantine tradition prevailing on a wide preventive action. As a matter of facts a certain amount of HIV infections, of TBC spreading, of endocarditis from Candida can be attributed to the lack of medical treatment of drug addicts. Only in the last month harm reduction philosophy has been introduced as a compromise in front a spreading epidemics, but in May 1989 the Minister of Health of Italy voted against a more liberal restriction to the syringes sale; reasons were political and ideological! Small pox has been eradicated simply because rational principle of interruption of the contagion chain in the endemic countries succeeded into W.H.O.'s Assembly against the old-fashioned and bureaucratic trend to rely upon passive measures such as certifications, stamps, evictions. We should wish that similar trend shall prevail in the international effort for AIDS and drug addictions containment.

## Introduction

Two different philosophies are colliding into the field of Public Health: the system of protect people through quarantine measures (in the last century enforced by "cordon sanitaire") and the trend towards personal and environmental prevention taking into account social and cultural causes of diseases and disabilities. The history of most past epidemics suggests that quarantine measures never succeeded in restricting the spread of cholera, while in many instances such as leper patients confinement favored the worsening of clinical cases and the spreading of contagion within the families restricted. In Guantanamo Naval Hospital now in the 1993 Autumn more than 300 Haitian discovered to be HIV positive are forced into a ramshackle camp with an equivalent number of armed soldiers to control. A common cited reason for continued incarceration of the Haitian refugees was that it would be fiscally irresponsible to take them into the United States when health care costs are skyrocketing and many American AIDS patient lack adequate care (1). But under the economical point of view the American tax-payers spend monthly around a million dollars to run the Guantanamo (Cuba Island) internment camp. Recently an interim order of a District Court Judge imposed to the Government to immediately move all HIV-infected detainees who had CD4 counts below 0.20 x 10 /L to a site where they could receive adequate medical care. In the meantime the overcrowding of barracks, the lack of sanitary facilities will provoked a general worsening of clinical conditions. The resurgence of Tuberculosis within HIV infected imposes on the contrary strict isolation measures in framework of the prevention of airborne diseases (2). Quarantines are more an expression of social distress and fear than an effective measures for controlling disease; it avoids the application of preventive measures such as environmental restoration (for cholera) immunizations, health education and interventions in order to modify negative behaviors and risky factors.

## Public Health and Politics

Public Health is not only a medical branch dealing with Hygiene, Prevention and Medical care because it comprises the social, cultural and legal sides and should develop its interventions within the boundaries of National Planning and Governmental policies.

There are not only budgetary restriction but also the need to manage to get into the general political framework of the Government (both National and Local). But in order to accomplish with advisory technical role all Public Health Officers should be able to produce objective advices taking into account only reliable data and avoiding any influence by ideological positions. When a famous American General Surgeon who was openly against abortion was asked for a report about the psychological consequences of abortion his report drawn-up after a careful consultation declared that non real danger of long-acting negative consequences in face of unwanted pregnancy could be quoted from the meta-analysis of the numerous studies performed (3).

In the same time Public Health Officers should act as health educators in the regard of the Ministries, Mayors, Politicians and Governmental Officers: all this people from one side can have the same "internal or external locus of control of his/her health" (4) typical of the social and cultural group they originate from, while they are in a peculiar risky situation in face of popular emotions, irrational panics and negative attitudes towards some diseases such as leper, cholera, mass foodborne out-breaks and, naturally, AIDS.

Positive attitudes cannot be "inoculated", but requires a long period of evaluation of the prevailing atmosphere and of gradual but constant and firm action of information, discussion, clearing-up of the various sites of the problem with some attempt to reach a Delphi agreement. When other influences are prevailing and the position and proposal of Public Health are shelved it is necessary to point out that the final decision is not the fruit of a technical advice.

Public Health Officers should be trained and selected in the network of Public Health Schools and the present overstressing of managerial "look" has to be integrated into a strong framework of biological background and firm commitment to a specific role (5).

As far as the relationship with the Bench is regarded it is necessary to establish an active interaction avoiding conflicts about the correct interpretation of the enormous amount of Laws and Regulations, taking into account the final objective of the Law and not the exact meaning of an isolate sentence. Generally speaking one point of database is represented by the trend of a rigid implementation of old-fashioned and ritualistic measures adopted for the sake of safety of administrators more than for a real protection of people and environment.

Food hygiene and food-handlers periodical controls are typical example of costly and useless measure adopted for the sake of safety of administrators more than for the real protection of people and environment. Food hygiene and food handlers periodical controls are typical example of costly and useless measures and the recent history of salmonellosis epidemic points out of our lack of real control measures at the primary source of the food-chain has been the main cause of the spreading, in spite of the numerous local investigations on personnel and premises. Hygienic education of food handlers should be the main action to reduce contamination at the production and at the delivery points, but such as measure is practically a theoretical formulation in face of periodical and useless stool examinations (6).

## Public Health and Drug Addiction

It is difficult to drawn a clear scenario regarding the optimal role of Public Health in the field of social and political decisions regarding drug abuse, Prohibitionism versus legalization and addiction as a social and not medical disease. The fashion of demedicalisation is a challenge for too many people, but, at least there are few clear point of understanding and of objective reasoning. When Public Health is facing chronic diseases whose aethiopathogenesis is still controversial the policy is to create a consensus giving room to all the opinions and prospecting pros e cons with a neutral attitude. It is not necessary to give priority to the medical labeling of drug addiction as a relapsing metabolic disease for understanding the biological principles of Methadone Maintenance clearly defined by Dole and Nyswander (7). In the same time under the point of usual allopathic medicine principles it is not possible to accept the thesis that it is wrong treating an addiction to a drug by substituting another addicting drug, some sort of "legalized addiction", a view that ignores the essential issue of patient outcomes.

There are so many other treatments provoking addiction (Corticosteroids, sleeping pills) and the research is aimed to substitute the addictive drugs not to abolish letting people without medical treatment. Even in countries where methadone historically was firmly established as Dole have pointed out (8) "Bureaucratic control of Methadone programs has given us "slots", a rule book and an army of inspectors, but relatively little rehabilitation". More recently a carefully study on various kind of methadone services in the U.S. (9) showed that methadone alone (even in substantial doses) may only be effective for a minority of eligible patients.

The addiction of basic counselling was associated with major increases in efficacy and the addiction of on-site professional services was even more effective. That was since the beginning the program of all the operators dealing with methadone maintenance, accused to be "State-drug dealers". In Italy there was and there is a lot of official statements picturing methadone as a "failure" with the support of anecdotical reviews, small clinical trials almost always performed with very low dosage. Regional Laws authorized methadone only for detoxification or, but with many requirements for long term treatment. Such a treatment should be performed with a maximum dosage of 35-40 mg. something against the same principle of blockade, apart from the arbitrary interference with the physicians right and duty to prescribe drug according to "Science and conscience". The present registration of Methadone allows for a single maximum dosage of 20 mg. which means 3-5 doses to drink in front of the operator when the maintenance is effective. Clearly everything is planned in order to utilize methadone practically only for the typical 21 days of detoxification, being on the market single lower dosages (10 and 5 mg. always in orange juice). A physicians prescribes methadone in a dosage more than that allowed by Regional Laws can be persecuted for illegal treatment with addictive substances.

Even after the victory of the Popular Referendum abolishing limitations to medical treatment of drug addiction in some highly civilized part of Italy (Prato-Toscana) the following "medical" order is still enforced (7 July 1993): (10) "Methadone treatment of drug addiction is comprised into a therapeutical Plan where physical detoxification should be accompanied by the more important psychic detoxification. According to such a priority Methadone treatment is limited to four weeks and cannot been repeated before two months after the last treatment. The second urine with drug methabolites causes the immediate interruption of the therapy; any exchange or manipulation of urine samples provokes the immediate interruption of the treatment: when the client is in probation, the Judge should be immediately informed. Any incorrect behavior in the regard of the personnel or when the Penal Code is violated causes the sudden interruption of treatment".

First of all there is no definition of "psychic detoxification" and in any case there is no knowledge of psycho-social treatment having effect after 4 weeks. Secondly how to justify under the medical point of view the deliberate provoking of an abstinence syndrome from methadone and the two months delay before another treatment?

The only reason is a "punishing trend" against a self-provoked disease: the physicians who work into the public network of Drug Addiction treatment apart from being public officers should remember to apply Helsinky Treaty for medical treatments and only the lack of a tradition of Public Health at National and Regional level can explain the enforcement of this kinds of medical orders.

The prevailing atmosphere of punishment is the opposite of the usual principles of medical treatment (to give assurance and support to the patients). In a very interesting research Rosenbaum (11) showed that among addicts addressed to Methadone Maintenance a pessimistic trends towards the value of the therapy was prevailing while a lot of somatic aliments were attributed to the drug (even osteopenia!). The same happens with overdose emergency treatment. The most part of deaths from opioid overdose are attributed to the morphine action in the respiratory system and a specific antidote or better antagonist, Naloxone is available. The Pharmacopea gives to the pharmacist the possibility to deliver Naloxone vials without medical prescription (another injectable substance included in such an exception is anti-snake serum). But, reading the official report on overdose deaths (considered as one marker of the epidemiology of drug addiction) no mention is made about the attempt to contrast morphine effect with the antagonist and only few services inform correctly clients about Naloxone and encourage the purchase or the distribution of at least two vials (cost of about 10 dollars) as a pocket first aid instrument. On the contrary all the booklets edit by Public Authorities for diabetics insist on the symptoms of hypoglicemia and suggest to bring always some sugar particularly when insulin is the main medicine.

Demedicalisation of drug addiction delayed the attention to the risk of entry viruses, germs and Candidae into the blood stream, while harm reduction intervention could have been performed many years before the AIDS epidemic i.e. in the 70thies.

## Public Health, Drug Addiction and AIDS

Emphasis encircling since the beginning Public Health's action in AIDS field was mainly supported by the death sentence applied without appeal and by the peculiar linkage with "Sin and Sex". The fact that in some country intravein drug addicts maintain the first position amongst AIDS cases did not provoke Public Health preventive intervention.

On the contrary in some country (i.e. Italy) penalties for drug possession were enforced by Law and number of addicts confined into jails increased, thus creating a medium for contagion through unsafe drug injecting (typical of all the prisons) and homosexual more or less forced activities.

Only after the last National Conference on Drug Addiction (Palermo 24-26 June 1993) and the result of a Referendum above quoted (April 18, 1993) the idea of harm reduction was introduced into the official vocabulary.

Maybe the stable door has been shut after the horses has gone (in the sense that a consistent part of addicts population was infected in the period 1982-89). Nevertheless it is important to regain the time lost and to overcome the negative attitude by part of decision-making centres, to win the prejudices and to offer to drug addicts both the best treatment available and the harm reduction possible, prevention of addiction remaining the first goal.

It is not possible to conclude without quoting an international phenomenon picturing the deep influence of political reasons on general decisions. During the yearly session of Ministers of Health of EEC Countries hold in Bruxelles on 16 May 1989 one of the many conclusions submitted was not published because of the negative vote of one

Country (Italy). The recommendation was entitled "Provision of safe injection material. The text was restored after another meeting (Rome December 1990) (12): It suggests:

- free delivery of needles and syringes (in some European country medical prescription is still required but not in Italy)
- or programs of needle exchange and of free distribution of disinfectants. Knowing the liberal legislation of Italy where syringes are available in the supermarket the other Ministers were rather upset; ambassadors too because it was the first time that a conclusion-recommendation born after long reciprocal consultation, official meeting and preventive submission to the ministries was not approved by a single Country (which means no publication on the Official Journal of the European Communities).

The explanation for the Italian rejection was that "the present Government supports a penalization of drug possession and owing to this trend cannot approve even a recommendation which gives legitimation to a criminal behavior"!

It is difficult to calculate how many drug addicts became HIV positive for lack of pharmacological treatment and for difficulties in purchasing syringes during the night, but in any case the above mentioned situation should give priority to the effect of building a real Public Health System, Philosophy and Power in Italy.

## References


1. Skolnick A. Doctors help HIV-Positive Haitian Refugees Gain Freedom From U.S. Government Detention Camp. J Am Med Assoc 270:563-564, 1993.
2. Bayer R, Neveloff Dubler N. The Dual Epidemics of Tuberculosis and AIDS: Ethical and Policy Issues in Screening and Treatment. Am J Pub Health 83:694-754, 1993.
3. Scotland N L. The Myth of the Abortion Trauma Syndrome. J Am Med Assoc 268:2075-2078, 1992.
4. Green L W, Lewis F M. Measurement and Evaluation in Health Education and Health Promotion, Mayfield Publishing Company, Palo Alto, CA, 1986.
5. Roemer M J. Higher Education for Public Health Leadership. Int J Health Serv 23:387-400, 1993.
6. W.H.O. . Health Surveillance and Management Procedures for Food Handling Personnel Techn, Rep. ser n. 785, Geneve, 1989.
7. Dole V P, Nyswander M E. Addiction: a Metabolic Disease. Arch Intern Med 120:19-24, 1967.
8. Dole V P. Hazard of process regulation: the example of methadone maintenance. J Am Med Assoc 267:1664-1668, 1992.
9. McLellan A T, Metzger P S, et al. . The Effects of Psychosocial Services in Substance Abuse Treatment. J Am Med Assoc 269:1953-1959, 1993.
10. Villani G. Norme per l'ammissione al trattamento con metadone, SerT USL 9, Regione Toscana, Prato, 1993.
11. On: Bollettino per le Farmacodipendenze e L'Alcoolismo anno XIII-1990 serie AIDS Vol.1 (Gennaio, Febbraio 1990) at page 124 only italian text paragraph C: Safer injecting material is published with a note (approved by 11 Countries except Italy).
12. On: Joint Drug Abuse Comprehensive Centre (D.A.C.C.): a publication for the first Pan-European Ministerial Conference on Illicit Drug Abuse Problems Oslo 9-10 May 1991 at page 159 among the European Plan to combat Drug Abuse Rome 13-14 December 1990: CELAD calls on the Member States to promote the measures referred to in part III C of those conclusions.

# 22
# The Dark and Light Aspects of Italian Legislation on Substances of Abuse

**A. Coluccia**
Department of Legal Medicine, University of Siena, Italy

## Summary

On June 1990 the Italian Parliament enforced a new law (Public Law 162) that considered illegal the use of any dose of illicit drugs. The approval of the law was preceded and accompanied by endless debates between political, humanitarian and scientific agencies taking side in favour or against the principle that personal use of an illicit drug is a crime.
The 162 Law introduced also the legal concept of "average daily dose" (i.e., the dose of an active substance sufficient to cover the daily needs of a drug addict) as a criterion for: a) distinguishing between a drug addict and a drug dealer; b) graduating the severity of the sanction (administrative for the drug addict, and penal for the drug dealer). Moreover, the drug addict has the choice between the administrative sanction and the acceptance of a therapeutic program. The commutation of the penalty will hold only if the therapeutic program results in a complete detoxification of the drug addict. In fact, relapse into drug use is equated to recidivism into legal offence and, as such, punished with penal sanctions. Such an ambivalent evaluation of a behavior, which can be either punished or cured according to rather arbitrary criteria, became matter of heated debates. The legal principle according to which an inadequate behavior can be considered pathological to the limited extent that a therapeutic prescription will prevent relapses, only applies to a few acute disease, and denies the existence of the most common chronic ones. The continuously increasing pressure of criticisms resulted in the call for a popular referendum which , in 1993, abrogated the principle of illegality for the personal use of illicit drugs. Hopefully, this fact may become the basis from which a rational bill for drug addiction control may arise.

Italian legislation regarding substances of abuse represents one of the most efficacious points of observation of the difficulties encountered by the state to legislate on drug addiction and, above all, on the personal use of drugs.

Political and personal concepts have often contributed in confusing the true requirement of an issue which should predominantly be of a purely technical nature. We must remember to distinguish, without forceful restraints, the figure of the drug addict "tout court" from the small drug pusher, the necessity of constructing a health system separated from a punitive one, which comes to the aid of the addict, and finally, the need to create centres of assistance for drug addicts without interfering with the "potestas curandi" of the physician.

In Italy we have assisted to an ensuing epistemological "querelle" on the matter, between medical and political knowledge, the former reaffirming a role through the dynamic experience that leads from diagnosis to treatment, the latter legislating according to unquestionable choices of politics: choices, which at times, bring to ignore the pharmacological data offered to the politics of social defence.

Law n° 162 of June 26, 1990, summarizes and is influenced, under certain aspects, by this "querelle", but also unfolds within its construct the difficulties of the State to legislate upon the most important issue concerning the legal framing of the personal use of substances of abuse and the consequent impediments in combining the health aspects with the repressive ones.

The first so called modern legislative intervention, in Italy, against the use and traffic of substances of abuse, is Law n° 685 of 1975 which is characterized by the consideration of the drug abuser as a patient and not as an offender.

In the eighties, especially in the late eighties, drug addiction begins to be considered as something which involves society as a whole, a concept which opposes the cultural position held in the seventies that the act of abusing drugs is something that belongs to private liturgies. The consequence of this new position is a crack in the substances of the preceding law, and that is, that drug abuse is a fact, situation or choice of individual.

Law n° 162 of June 26, 1990 contains two conceptual poles, and these are, on the one hand, that drug addiction is an illegal activity which is, within the limits of the law, left to individual management (let us recall the guarantee offered by the article 80 of the preceding 1975 law allowed for the possession of a "moderate quantity" of substance of abuse for personal use) and on the other, that activity of drug abuse is a situation which calls for the evaluation parameters of collective interest. This law verges to repression when it affirms that drug abuse is a legal offence, even if law n° 162/90 tries to mitigate the choice creating a series of "health opportunities" (art. 75 and 76 T.U.n° 309/90) for the addict rather than enforcing the administrative and penal sanctions.

In the 1990 Law the penal sanction (or administrative in the prefecture rite) and the therapeutic "chance" live together harmoniously, and the activation of the restrained subject or on the basis of the legal role invoked, the marked, accused, or condemned subject, will activate either the road to punishment or to cure. However, the formulators of this norm, knew full well that the difficulties within the association of punishment diverse legal titles and treatment would represent the weak point of law, especially when considering the ambiguity of the role played by those public or private (e.g. therapeutic communities) structures in aiding the drug abuser.

The Prefect is the first legal figure encountered by the addict that has been restrained for the personal use of substances of abuse. The Prefect's competence regards only that person which has taken a dose of a drug that is not superior to the average daily dose established by decree n° 186, December 19, 1990 of the Ministry of Health (decree which considers the quantity of active substance within the concept of average daily dose).

The so-called prefectural rite ex-art.75 outlines several situations based upon the use of heavy or light drugs and the subject's age (minor or legal adult).

In the hypothesis that the individual has used the so-called light drugs indicated in

Tables I and II and IV (Tables which contain light substances considered substances of abuse), the Prefect may issue an administrative sanction for a period from 1 to 3 months and that is suspension of: a) drivers license; b) passport; c) fire-arm license or he may define the proceeding by formally inviting the subject not to use drugs.

In the case of personal use of heavy drugs (Tables I and III) the administrative sanctions are applied from the beginning, for a period which ranges from 2 to 4 months, or the individual may choose to enter a therapeutic program.

However, 2 defaults of the administrative sanctions or the refusal of the therapeutic program leads to the praetorial rite of art. 76 (T.U. n° 309/90).

When the restrained subject is a minor (less than 18 years of age according to Italian Law), the Prefect has the possibility of defining the proceedings by "formally inviting the individual not to use substances of abuse" regardless of the substance used. In this manner, it seems that the legislator has tried to free the "deviated" minor, from the stigma of "Junky" and from all the negative correlations associated to this image.

If the minor is restrained once again, the eighth clause of art.75, states that the Prefect may "if possible and considered opportune" ask the parents to intervene and instruct them on the social services structures present within the territory.

These conditions have created a great deal of perplexity: first of all, the fact that young people are at the highest risk of drug abuse should have foreseen for family involvement right for the beginning: it's inopportune, especially from the therapeutic point of view, for the family to be absent at the time of the first restrain. Furthermore, the formal invitation of the Prefect to not make any more use of drugs seems more to satisfy a paternalistic role rather than the need of affronting the problem in its true context, and that is a different situation which has for protagonist the minor, his problems and his family.

From the third violation on of art. 75, adults (for minors the competency is held by the Court For Minors), regardless of the substance of abuse used, are subject to legal proceedings (ex art. 76 T.U. 209/90).

However, in the case the minor has respected the administrative sanctions or has successfully entered and continued a therapeutic program, the Prefect legally closes the case.

The above-mentioned procedure, but above all the insertion of average daily dose as a point of division between a situation which is always illegal and suspended, on the base of the quantity of average daily dose, between administrative or legal sanctions, and the cure as a "catharsis" of the illegal substances consumed, and furthermore, the quantization of the average daily dose, create strong criticisms toward the legislative asset inaugurated by law n° 162/90. Moreover, many from the health profession protest against this law, because they feel deprived of their medical role by the dictate of the norm, since the art. 83 considers as a crime the prescription by a physician or veterinarian of substances of abuse or psychotherapy (art. 83 law n° 162/90 substances inflicts the same punishment describe by art. 73, clause I, for non-therapeutic use 4, 5; traffic, production, etc.).

Even where the treatment of a drug abuser is specified by the law, the physician's activity is highly restricted since he must signal out the addict to the SERT (art. 121), thus objectively losing his "potestas curandi" with such an act.

Once the drug abuser has been signalled out (ex art.121), the SERT physician takes control on the health situation of the drug abuser. A further restriction enforced the individual's physician and also on the SERT physicians is the legal imposition by a Ministry Decree of the drugs which can bc uscd as substitutes for the treatment of drug addiction.

Indeed, on the basis of the Decree n° 445 by the Ministry of Health of December 19, 1990, the only drug allowed as a substitutive measure is chlorydrated methadone in syrup (ex art. 1 D.M. 1990/445) which can be prescribed only by SERT physicians or "appropriate hospital or university structures".

Numerous criticisms are raised against the law of 1990 which did not organize nor diffuse on a national level the appropriate structures, centres of aid for drug abuser. However, the strongest condemnations are raised against two points:

a) making the personal use of substances of abuse a crime, and the consequent "state of accused" of the average daily dose, which through the use of the tabular system (ex Ministry Decree n° 187) reduces individual dependence on substances of abuse to precise parameters fixed by the law;
b) removing the power to cure from the health professional since the decree of the Ministry of Health (December 19, 1990 n° 445) regarding the use of drug substitutes in the treatment of addiction, imposes the use of methadone as the only drug, and the administration of this drug only by SERT physicians.

For these as well as other reasons, Italy will go through a much labored course of realization of this law. Indeed, there wont be a lack of appeals to the Supreme Court that will ask for the unconstitutionality of the law 2. However, thanks to the Antiprohibition League with its popular Referendum of April 18, 1993 the most discussed articles will be abrogated, and thus Italian legislation on the matter will be, at least in part, rewritten.

The popular Referendum has abrogated the controversial notion of average daily dose (ex art. 75 T.U. n° 309/1990), therefore there is no penalization for the detention of substances of abuse (without limits of quantity) for personal use, with the consequence that whoever imports illegally, or acquires or detains substances of abuse or psychotropic drugs for personal use is subject to administrative sanctions under the competence of the Prefect. The praetorial rite ex art. 76 responsible for penal proceedings for the drug abuser is also abrogated (for those subjects that have violated the administrative sanctions twice).

The referendum of April 18, 1993 restores a decisively different dimension from that defined by the 1990 Law to the drug user, since the use of drugs loses its penal significance and acquires the personal aspect of choice, subject only to administrative sanctions. However, the most important point is the abrogation of the D.M. n° 445 of December 19, 1990, specifically the norm contained in art. 2, clause 1, letter and n° 4 which conferred the power of determining the "limits and modalities of use of the substitute drugs" to the Ministry of Health (Furthermore, the Decrees by the Ministry of Health of August 7, 1980 and October 11, 1980 approved by the D.M. of Health n. 445/90, concerning the regulation of substitute drugs in the treatment programs for addicts has now been abrogated).

Thus, the power to cure requires that dimension which has been traditionally defined by medical science. Although, a "memo" of July 30, 1993 from the Ministry of Health invites the Counsellors of Health of all Regions and autonomous provinces to consider when evaluating the clinical case of the individual drug abuser, the opportunity of sending the individual to qualified public structures for treatment, since the treatment is complex and long-term and requires the appropriate programs correlated to the gravity of the associated diseases. The memo ignores the contents of referendum and reaffirms "de quo" the sole use of chlorhydrated methadone in syrup form as a substitute in the treatment of drug abuse and entrusts the SERT once again the "responsibility of defining a program for the execution of a therapeutic treatment with substitute drugs indicating the modalities, dosages, duration, and control criteria...". We feel, regardless of whatever ideological option, that the two norms do not hold their legal reference within the law 162/1990 since art. 2 clause 1 n°4 ("limits and modalities of substitute drugs") has been abrogated, and that this has also led to the abrogation of D.M. 445/ 1990. Thus, we believe that there should be no restriction on the pharmacological prescription of the curing physician.

Furthermore, in view of the abrogation of clause 1 of art. 121 (which imposed the physician to signal out the drug abuser to the SERT) the complete competence of the physician, in general, must be applied and imposed to the treatment of the drug addict with substitute drugs, and to the action and role of the SERT. Due to the elimination of the limits on treatment and use of substitute drugs stated by the law of 1990, we will have, if no other legislative adjustments occur in the future, that the "physician has the right to prescribe whatever substitute to heroin, including morphine, in the quantity he feels most opportune, even for an undetermined amount of time (the so-called "maintenance treatment"). Furthermore, with the eradication of the discipline of modality, the taking of a drug can be left to the free choice of the patient without the need of controls" (3).

## References

1. The decree of the Ministry of Health of August 7, 1980 and on October 1980. Were also effective at the time.
2. Coluccia A., Marzi A. : La tossicodipendenza fra problemi vecchi e nuovi. Considerazioni medico legali e criminologiche in margine alla sentenza della Corte Costituzionale (10-11 Luglio 1991, n° 333) in Zacchia, 4, 1991.
3. Riccioni R., Riccioni M.M.: Gli stupefacenti. Cedam. Padova , 1993

# 23
# Methadone and Criminality. Euro-methwork Workshop

**E. Buning, P. Lang, M. Torrens, C. Herzog, L. Grönbladh, H. Joseph, A. Tagliamonte, O. Blix**
Euro-methwork, GG&GD, 1018 WT Amsterdam, Holland

The workshop Methadone and Criminality was chaired by Peter Lang from Bremen (Germany). After a short introduction about the history and activities of Euro-Methwork, four experts gave short introductions about the role of methadone in preventing criminality and about the use of methadone in the prison system. These introductions were followed by a discussion.

## Introduction

Ernst Buning, coordinator of Euro-Methwork, presented the bringing about of Euro-Methwork: A group of European methadone experts met in Orlando, USA, at the American Methadone Conference in November 1992. They were impressed by the way the American methadone experts were communicating with each other and realized that they, as Europeans, often don't even know what other cities in their own country were doing, so let alone knowing what is happening in other European countries. They decided to organize some sort of a network.

The French organization 'FIRST' took the initiative for further action. Ann Coppel of FIRST and Ernst Buning of the Amsterdam GG&GD went to Brussels and met with the team of the European community who is responsible for setting up the European Drug Monitoring Centre. They asked their support. As a result, the European Community provided some money and Euro-Methwork could start its activities.

The main goal of Euro-Methwork is to bring people together and to facilitate communication between experts working in the methadone field. It concerns researchers and policy makers, and especially persons working in methadone programs or involved in setting up methadone programs. They are brought together to see what can be learned from each other, so that they don't make the same mistakes that others made when setting up a methadone program for the first time.

To facilitate the communication, a database has been developed and Newsletters are produced and sent around. Participation is free of charge and no special qualifications are needed: anybody can be part of Euro-Methwork.

One of the activities was the organization of a number of workshops. These are low-costs workshops since they were organized at conferences and seminars, whenever people are together anyway and participants have already paid for their travel expenses, hotel etc.

One workshop was held in Rotterdam during the Harm Reduction Conference in March 1993. This workshop was video-taped by a professional crew. This video-tape is entitled "Methadone in Europe". On this video, experts from methadone programs in Europe discuss their programs and their experiences and present research finding, obstacles in setting up methadone programs and strategies for overcoming such obstacles.

The second workshop was held in Berlin during the World Aids Conference. There, the discussion specifically targeted at the relationship between methadone and Aids. In Berlin, about one hundred people came together, talking -from different perspectives- about the relationship between methadone and Aids.

In this third workshop the relationship between methadone and criminality will be discussed.

## Controlled Programs

Leif Grönbladh presented the methadone program in Uppsala, Sweden. This program is based on the original ideas of Dole and Nyswander. In their study, Leif Grönbladh and colleagues looked at 345 clients who were followed up for at least 5 years. In some cases they were even followed up for 20 years. The outcome very much resembles that of the methadone programs in New York in the early seventies. Amongst the patients who stay in the program there is a dramatic increase in the number of people who started to work. Also the use of illegal drugs decreased and the level of criminality dropped drastically. In the program, the team tries to create an atmosphere of dignity, which is essential if the patient is to overcome his addicted life-style. They consider methadone treatment as a long term process. According to Leif Grönbladh, crime reduction will only occur in comprehensive methadone programs. "Harm-reduction and legalization don't reduce criminality in the same magnitude as a well controlled methadone program".

## Low-Threshold

Marta Torrens discussed the preliminary results of a study about the quality of life of patients in a methadone program in Barcelona. In her presentation she described a six-months follow up of 30 patients who were in methadone maintenance for the first time in their lives. Both the Nottingham Health Profile (NHP) and the Self-Reported Health Scale (SRHS) were used at baseline and six months later. The methadone program in question is a low-threshold program with minimum intake criteria. Priority is given to patients with Aids, tuberculosis or psychiatric disorders and pregnant women. The program has no time restrictions and there is no maximum dose. Clients have to come daily. Only in cases of violence, disrupted behaviour and/or drug trafficking in the centre, can clients be dismissed from the program. Clients who are still using illegal drugs (positive urine) are denied certain privileges.

Among the group that was followed for 6 months, there was a significant drop in the number of arrests by the police. After 6 months of methadone treatment, 70 percent of the clients had ceased using illegal drugs, HIV risk behaviour had decreased, medical control was intensified and a general improvement of quality of life was observed.

**Rykers Island**

Herman Joseph from New York explained how hepatitis and drug related crime decreased in New York city after 20,000 heroin addicts enrolled in methadone programs in the early seventies. There are about 200,000 heroin addicts in New York. The criminal Justice System is a focal point for addiction. In the late sixties, after riots had taken place in prisons because addicts were suffering severely after their arrest, Vincent Dole set up a model methadone detox in satellite prisons throughout New York. Nowadays, addicted prisoners are sent to Rykers Island Central Jail Facility, where they are detoxified. About 5 years ago, methadone maintenance programs were set up in the jails, as a result of the large number of drug users with Aids. Although there was considerable resistance among prison officials, the first methadone maintenance program was set up in the Women's Prison. This was very successful and was therefore extended to the men's houses. The anticipated problems such as discipline and diversion did not materialize. The contrary happened: prisoners were easier to handle and it had a positive effect on the morale of the prisoners.

**Referrals**

The methadone programs in the NY-jail made connections with the methadone programs in the city. A problem in referring ex-prisoners to these clinics is that many ex-prisoners come from the lower socio-economic strata, often homeless, unemployed and don't have the appropriate papers for entrance to methadone clinics. Furthermore, they may not have social benefits and insurance. Prisoners are sometimes released in the early hours of the morning if their time in prison is up. When they are homeless, they may immediately relapse into their addictive behaviour and will not go to the methadone clinic. Despite the fact that there are problems in creating a smooth transition from the jail's methadone program to the City methadone program, 30% of the referrals is successful. Another problem is that of those patients who are living in such terrible situations (homeless, jobless etc.) and who will commit a crime just to get back to prison.

## Basel Prisons

Christian Herzog from Basel, Switzerland, focused on the use of methadone in prison to prevent the further spread of Aids. Most of the drug addicts in Swiss prisons have a long history of drug use. The sero-prevalence ranges from 35 to 40%. Since drugs are used in prisons and needles are shared, prisons can be a place for further spreading HIV in the drug using community. The city of Basel runs two prisons with approximately 200 places. The prisoners detained are either in custody for a few days or weeks or in prisons for a "short" sentence of less than a year. In January 1991, methadone substitution was introduced for drug addicted prisoners who were already on methadone when entering the prison.

When the methadone program started, urine checks showed that all sorts of legal and illegal drugs were used before and during detention. During their stay in prison, those prisoners on methadone drastically reduced their use of illegal drugs. The evaluation of this methadone program is very positive. However, Christian Herzog has one critical comment: "During the two years, a number of disciplinary detoxifications took place. These episodes are in my opinion neither ethically nor medically justified and they are an offence against human rights".

The future of interventions in the Basel prisons has recently been formulated by a commission. This commission came to the conclusion that drug users in prison should have the same treatment options as outside the prison system, i.e. (1) detoxification, (2) methadone substitution and (3) access to sterile needles and syringes.

## Detoxification

A doctor who used to work in a prison in Geneva told the audience that he was only allowed to do short detoxifications with methadone (maximum of 10 days). Now the situation is different. If an incarcerated drug user participates in a methadone program, he will receive methadone whilst in prison. However, the maximum dosage is 40 milligrams per day. So if he was on 100 milligrams, this is reduced during imprisonment to 40 mg per day.

Dr. Tagliamonte from Italy, who has a 10 years experience working in prisons with drug users, explained his interventions. Addicted prisoners who were not in a program received a dosage which would be increased the first couple of days until they feel o.k. Those who were already on methadone, received their usual dosage for the first three days. Subsequently, all prisoners on methadone would be slowly detoxified (5 mg per three days). During the 10 years he has detoxified about 1,000 prisoners. In his experience a detoxification to 20 mg is no problem. But once arriving at 20 mg the problems begin. Sometimes it was necessary to slow down the detoxification. Fifteen days before leaving the prison, those who are clients of methadone programs, can receive methadone again.

**Ideology**

Olof Blix explained that in Sweden, all methadone patients in prisons are detoxified. He stressed the possibilities of making drug users drug free whilst in prison, since they would have better possibilities to remain drug free after their release from prison. This statement provoked reactions from the participants of the workshop. "Ideology" someone shouted from the back benches. In a reaction to this, someone from the audience stipulated the fact that at least in Sweden a drug user could receive methadone, whilst in countries such as Greece and France methadone is not or hardly available.

**Differences**

"Nevertheless," someone else added, "it seems unfair that there are such major differences between the various EC member states. Suppose we had the same differences in the treatment of diabetes. That in one country doctors would deny insulin to diabetic prisoners because they don't believe in it, whilst in another country the diabetic prisoner would receive an adequate dose of insulin. Wouldn't we all feel that this would be ridiculous?"

In Norway, the situation is very different from Switzerland. Since only few drug addicts receive methadone in Norway, so far only four methadone patients have been in prison. "For them it was impossible to arrange for continuation of their methadone provision" stated Gabrielle Welle-Strand, who runs the Norwegian methadone program.

**Conclusion**

Evidence from both a strict methadone program in Uppsala and a more relaxed "low-threshold" methadone program in Barcelona, show the effect of methadone in reducing criminality. Correct dosages (somewhere between 50 and 100 mg per day) may be the key in reducing criminality. It is a reality that illegal drugs are available in most prisons in the Western world. The experiences with methadone in prisons in Basel and New York are positive. Provision of methadone in these prisons reduced the use of illegal drugs and ameliorated the general climate in the concerned prisons.

Anybody who is interested in Euro-methwork should write to:

Euro-methwork
GG&GD, room 510
Nieuwe Achtergracht 100
1018 WT Amsterdam, Holland

tel: (31 20) 55.55.307
fax: (31 20) 55.55.775

# 24
# Workshop of the European Collaborating Centres in Addiction Studies (ECCAS)

A selection of papers was presented at the above workshop. The purpose of the workshop was to introduce ECCAS, present the preliminary findings of its first major study and present short papers on some of the work and the experiences of the participating centres.

## The ECCAS Network

Established in April 1992, ECCAS (European Collaborating Centres in Addiction Studies) is a developing network of professionals committed to advancing current knowledge in the field of substance misuse, which should have an impact at all levels: clinical, research and policy. The Network currently consists of sixteen centres in fourteen European countries: twelve member states of the European Union, Poland and Switzerland.

Since its inception ECCAS has been co-ordinating a large European Study, in which thirteen of the sixteen centres are taking part , investigating methadone substitution therapy and its impact on HIV risk behaviours; the background and initial findings are presented later in this paper. Apart from co-ordinating the largest study of this nature ever undertaken in Europe, ECCAS is also involved in a variety of other activities. Although at an early stage in its development, ECCAS produces a bi-monthly newsletter, and has plans to expand its program to include education and training initiatives, staff exchange programs, to co-ordinate and facilitate conferences on all matters related to the addictions, and to continue with its collaborative research program. With respect to the latter, ECCAS has already been involved in developing several research protocols, including funding applications.

ECCAS provides a much needed network and forum for concerned professionals either directly or indirectly related to the field of substance misuse, to communicate, collaborate and share relevant experience, knowledge and skills in an ever changing and extremely challenging field.

**Methadone Substitution Therapy and Its Impact on HIV Risk Behaviours**
A Multi-Centre Study Co-ordinated by and Involving Thirteen Centres from the European Collaborating Centres in Addiction Studies (ECCAS)

**A.H. Ghodse, C. Clancy, M. Jones (London); M. Bourquin (Geneva); B. De Las Heras, M. Montes (Oviedo); E. Keenan (Dublin); I. Lopes (Porto); M. Riglietta, L. Tidone (Bergamo); C. Rosinger (ESSEN); K. Runge Nielson, (Arhus); L. San (Barcelona); F. Schifano (Padova); P. Schoemann, J. Werenne, L. Cremmers (Brussels); B. Sommer (Ringkobing); S. Wieviorka (Paris)**

## Introduction

In the last decade drug policies and treatment services have had to be reassessed with emphasis now being placed on the prevention of HIV infection; its spread is seen as greatly increasing the risks and harms of injecting drug use both to the individual and to society (1).

It has been claimed that Methadone Substitution Therapy (MST) helps to fulfil these aims whilst also acting as a "normalizer", allowing "recovering" opiate dependent individuals the opportunity to adjust to societal requirements (2). Further research is required, however, in this area to establish the effectiveness of MST both in the treatment of opiate dependent individuals and in the prevention of the spread of HIV.

The majority of evaluation studies concerning treatment of heroin dependence come from North America and, due to the differences in culture and drug legislation, they cannot be generalized to the European context. It is true too that there are wide differences in treatment regimes on offer for opiate dependent individuals throughout European countries. Many professionals in the addiction field throughout Europe are concerned about the possible differential impact of these policies, though no scientific evidence exists to support these concerns.

A European multicentre study is being carried out by the European Collaborating Centres in Addiction Studies (ECCAS) to develop research instruments to evaluate the impact of methadone substitution therapy on HIV risk behaviours. This study commenced in April 1991 with funding from the European Commission.

The efficacy of a treatment intervention can normally be determined by using clinical trials. Subjects randomly allocated to "treatment" versus "no treatment" groups can enable the measurement of differences in outcome. Because of the nature of treatment of drug dependent individuals it is not possible to use such a design. Apart from the ethical considerations, there is a practical problem in that drug users who are offered "no treatment" are unlikely to remain in contact with treatment centres. Furthermore a clinical trials paradigm is clearly inappropriate in a study which evaluates existing health service delivery.

A suitable paradigm was therefore designed specifically for this study. This involves the recruitment of 2 cohorts of subjects.

Cohort 1 consists of subjects who are currently in treatment, while Cohort 2 is made up of subjects who are newly admitted to treatment. While Cohort 1 can not be considered a control group, it acts as a comparison group. Thus any differences between the two cohorts at the baseline may disappear if MST is effective.

In order to be included into the study, patients must conform to the following selection criteria:

1) Fulfillment of DSM lll criteria for opiate dependence.
2) No clinical evidence of functional psychosis or chronic organic brain syndrome.

Structured questionnaires were used to collect the following information relating to a period of 3 months prior to interview:

Demographic data
History of drug and alcohol misuse, complications, injecting practices
Family pathology
Personal history including work record, criminal activity, sexual behaviour
Clinical investigations, eg. HIV test, Hepatitis B, C virus, liver function test, urine testing
Treatment previously received
Current treatment plans including patient's perception of drug and alcohol use, physical and mental state
Patient's attitude to drug injecting behaviour, safer sex, etc

The following variables were measured using objective indicators:

Health: general physical/psychological (time spent in non-drug-related institutions), HIV status (blood test)
Drug/alcohol use (random urine testing)
Legal (number of arrests/convictions, imprisonment, etc)
Employment (documentary evidence)
Interpersonal (objective reports from therapists)

There are two hypotheses inherent in this study. The first states that there will be a demonstrated effect on the physical and psychosocial features of opiate dependent individuals treated with MST. The second states that there will be some common features in this demonstrated effect across the different treatment centres throughout Europe. It was accepted that due to the ambitious nature of this study the project would have to be carried out in stages.

**Phase One**

During phase one of the study, thirteen centres were recruited, each agreeing to participate in recruiting a maximum of 50 patients into each cohort. Three questionnaires were developed, piloted, and adjusted to take account of national idiosyncrasies; the European Substance Use Database (EuroSUD), the Substance Abuse Assessment Questionnaire (SAAQ) (3) and the HIV Risk Behaviour Questionnaire (HIVRBQ) (4). These were subsequently translated into six European languages and data collection commenced in June 1992.

An extensive survey of each of the participating centres was also conducted. The Treatment Centres Characteristics Schedule was used to systematically collect data to provide a detailed picture of each participating treatment centre including information about physical facilities, operational policies, staffing, administration, counselling services, toxicology and methadone prescribing, medical services, record keeping and funding arrangements. The resulting information was essential to provide a basis for comparison between the centres in each country. It also forms a very important study in its own right (5).

### Phase Two

The development of reliable, valid assessment and research instruments, which may be used to evaluate treatment outcome in a number of settings throughout Europe, is seen as an essential component of this project. While these questionnaires had been fully piloted in Phase One of the study, it was necessary to carry out a detailed evaluation of the questionnaires once each centre had collected sufficient data to have become familiar with the tools. An extensive four stage evaluation process was therefore designed and carried out.

### Initial Findings

At this stage of the study it is not possible to present the findings in their totality. Base line data collection is on going and the centres are just commencing the follow up phase. However it appears that about a third of the study population are female, with heroin as the main drug of use followed by methadone. Over 90% of the study sample have injected drugs at some point in their lives and females subjects are one and a half times more likely to have a drug using partner than males. Differences in composition of the centres, as well as between the two cohorts are emerging.

### Conclusion

This study, and the networking activities of ECCAS is important not only in terms of the information gathered but also because it can contribute to the knowledge base for policy formation in Member States. The amount of work achieved in the last eighteen months, given that financial resources were limited, is an indication of the commitment within the European community to research and collaboration, at all levels, and although this study is at a very early stage, with further data collection and analysis necessary, the dissemination of information about the research instruments that has already been activated is of the greatest value. The proposed network activities will ensure that knowledge of ECCAS and the common assessment and research instruments which have been developed, along with expertise in applying the instruments in different treatment settings, will spread across Europe and even further afield.

### References

1. Advisory Council on the Misuse of Drugs (1993) AIDS and Drug Misuse Update (London HMSO).
2. van de Wijngaart, G.F. (1990) Supply on Demand: Maintenance or Abstinence. In Ghodse, A.H., Kaplan,C.D. and Mann, R.D. (eds) Drug Misuse and Dependence. Parthenon, Carnforth.
3. Ghodse,A.H. (1989) Drugs and Addictive Behaviour: A Guide toTreatment. Blackwell.
4. Ghodse, A.H., Tregenza, G. and Li, M. (1987) Effect of fear of AIDS of sharing injection equipment among drug abusers. British Medical Journal, 689-699.
5. Ghodse, A.H. (1992) Treatment Characteristics Schedule, report prepared and submitted to the European Commission. Division of Addictive Behaviour, St. George's Hospital Medical School, London.

## A Report from Denmark

**B. Sommer**
Embedslaege Institutionen, For Ringkobing AMT.
Norredige 20, 6950 Ringkobing, Denmark

Embedslaege Institutionen for Ringkobing County joined the ECCAS study in May 1992. At that time an internal review of the relevance and effectiveness of methadone substitution therapy was about to take place. Methadone had been used for almost 15 years in the county with these two issues constantly being debated.

The small number of drug addicts being treated with methadone in the county made it difficult to use traditional evaluation methods. The number on methadone substitution therapy was at any time approximately 20. Comparing our data with other counties or centres in- and outside Denmark was considered necessary if the study was to be reliable and useful.

A junior doctor was employed for the specific task of collecting the data. He is not a member of the staff and did not know the clients in advance. In October 1993 the second phase of follow-up interviews is about to commence.

**What Have We Learnt?** None of the sample group tested for HIV proved positive. Although one patient within the sample group tested positive for HIV initially this proved to be a false positive on subsequent testing. Testing for HIV in Denmark is only carried out on request from the client, so not all are tested, but the prevalence of HIV among drug addicts in Ringkobing county is presumably very low.

Due to the HIV positive status in other ECCAS centres, the Denmark response cannot afford to be complacent, but accept that this is a further indication for the need for closer co-operation in the future.

**What Do We Still Want to Know?** We are not much closer to any conclusion regarding the cost-effectiveness of methadone substitution therapy in Ringkobing county. As the risk of HIV transmission is still thankfully low, the question we are asking is, are funds and manpower resources better used on other social and medical programs? Hopefully the ECCAS study investigating the Evaluation of Methadone Substitution Therapy and Its Impact on HIV Risk Behaviours will begin to give us the answers.

## A Report from Germany

**C. Rosinger**
Klinik Fur Allgemeine Psychiatrie,
Hufeland Strasse 55, 45147 Essen, Germany

In the Federal Republic of Germany (FRG) three areas have been identified which indicate the need for methadone substitution therapy: for analgesic purposes, severe (somatic) comorbidity to the addiction (including pregnancy) and for the treatment of opiate addiction itself. The latter, as yet, is not covered by health insurance.

In the so-called old federal states of Germany the estimated number of those dependent on "hard" drugs is 120,000 - 150,000. Although 15% are considered to need methadone substitution therapy, only 4% actually receive this form of treatment from approximately 1,680 prescribing doctors.

In 1987 the federal state of Northrhein-Westfalia started a methadone research-project which was a pilot study on the feasibility of methadone substitution therapy as a treatment response to opiate addiction. This was a prospective and muticentre design with a high admission threshold (compared with long term abstinence programs). 247 patients were recruited during the period of evaluation (1987 -1992), (PROGNOS 1993).

The therapeutic aims of this study, whilst working towards abstinence from methadone, were to stabilize the individuals' problems in the following areas:

psychiatric
physical
psychological
social
drug use

The HIV positive status amongst the sample group was approximately 12% at beginning of the study (approximately 15-30% of drug users are HIV positive in FRG). During the observation period the index of seroconversion was approximately 0.5 per 100 patient years, which is a low number in the context of International data (0.9-7.9). The leukocyte helper cells were found to be stable, and the IGM-ratio which increases in function of frequency of intravenous use in our study indicated a decrease of intravenous injecting in the sample population.

The interest of the FRG in participating in the ECCAS study is to have an opportunity to compare the German results with the international data and further, to connect this data with other (projected) studies e.g. a national multi-centre-study concerning the "specific indications for different forms of therapy" and with other studies proposed and to be co-ordinated by ECCAS e.g. "service needs and evaluation of outcome of interventions in drug and alcohol misusers" or "nature and extent of substance misuse-development of an European database.

## A Report from Ireland

**E. Keenan**
Drug Treatment Centre
Trinity Court 30/31 Pearse Street, Dublin 2, Ireland

The National Drug Treatment Centre in Ireland has been in existence since 1969 and in its present location since 1988. It is involved in the assessment and treatment of patients suffering from all forms of substance misuse problems. The approach is a multidisciplinary one with a team consisting of doctors, socials workers, nurses, psychologist, occupational therapist, group therapist and community outreach worker. There is also a Health Education Sister whose work includes dealing with telephone advice calls (Helpline) and giving lectures in the community.

Programs offered for opiate misusers can be either methadone detoxification, methadone maintenance and a low dose methadone maintenance which runs in the evening time in conjunction with a needle exchange program.

All patients on methadone maintenance are involved in both individual counselling and group therapy. A separate group is in operation for the younger drug misuser, particularly those misusing ecstasy and cannabis and another for benzodiazepine dependence.

A special emphasis is placed on the pregnant opiate misuser with a program consisting of methadone maintenance throughout pregnancy and regular ante-natal check ups at the Centre from a visiting gynaecologist who attends once fortnightly.

All toxicology (via supervised urinalysis) is performed on site. There is a ten-bedded in-patient unit in a general hospital (Beaumont) where patients may be detoxified or stabilized on maintenance.

Patients attending the Centre are offered medical check ups and the opportunity to avail of HIV testing after appropriate pre-test counselling. Last year (1992) the Centre treated 1505 patients, an increase of approximately 25% over 1991. This included 663 new patients and 842 return patients who had in all 44,949 attendances.

An interesting trend in opiate misuse over the last seven years has been the emergence of prescribed opiates as major drugs of misuse. These drugs include Morphine Sulfate, Buprenorphine, Dihydrocodeine tartrate and methadone. In fact, in 1992 more patients presented to the Centre misusing MST - 856 - as compared to those misusing illicit Heroin - 836.

The Centre continues to expand and there are plans in 1994 to commence a rehabilitation Day Program particularly for younger patients who have completed detoxification. Also we intend to reintroduce a Basic Skills Retraining Program which has been successful for patients on methadone maintenance in the past.

Involvement with the on-going ECCAS study has been very beneficial for the Centre both from the point of view of establishing contacts throughout Europe and also to establish standards in the treatment of this difficult area.

**A Report from Italy**

***Methadone Treatment in Opiate Addicts: 12 Years Experience in Italy***

**M. Riglietta, R. Bagnato, M. Campana, P. Colleoni, P. Drago, M. Fusini, Z. Vyhna'lkova', L. Tidone**
SER.T. U.S.S.L 29 - Bergamo, Italy

The Addiction Treatment Unit of Bergamo Local Health Area, is an outpatient centre with a catchment area of 250,000 inhabitants. Methadone Substitution Therapy (MST) has been used by the Unit since 1980 after receiving ministerial approval. Patients filling the DSM III-R criteria for opiate addiction are commenced on MST and diagnosis is made following a detailed assessment procedure covering:

medical and psycho-social assessment.
medical examination.
general laboratory investigations and in particular tests for syphilis, hepatitis B , C and HIV.
urinalysis for the most common substances of abuse.

Following assessment, the staff team discusses and draws up a plan of treatment appropriate to the needs of the patient. Individualized goals are set with the patients, identifying the areas in their life which need most attention, e.g. medical problems, substance use, psychological etc.

We use methadone in three different ways:

1 Detoxification. The starting dose is between 40-70mg/daily which is then reduced over a period variable from 10 and 30 days.
2) Therapeutic Contract. Treatment is continuous for not less than three months and the aim is, after a period of stabilization (usually one month), the patient must be drug free from illegal opiates. Following stabilization should the patient subsequently submit more than three urine samples positive for heroin the treatment is discontinued.
3) Methadone Maintenance. Treatment continues despite "dirty urines" though there must be some evidence of a reduction in the individual's heroin use.

In 1992 the Unit has been in contact with 1100 drug users of whom 900 were opiate addicts; 700 carried out a specific treatment for addiction and 250 were methadone treatments. The mean dose of methadone is 50mg/day, with a range between 10mgs and 130mgs. In two different punctual relevations we found 30% of "dirty urines" for heroin and 5% for other substances. 60% of patients who had been receiving MST were also involved in another programs such as psychotherapy, self help groups and rehabilitation.

Our Addiction Treatment Unit started screening drug addicts for HIV infection in 1985. In 1992 the seroprevalence for IVDU was 24%, however during the period 1985 -1992 the following has been noted:

1) The prevalence of HIV positivity in the IVDUs that were tested for the first time in 1985 has had a negative trend, from 30% to 15.7% in 1992.
2) In the 7 years ('85-'92) we have seen 19 seroconversions. Whilst 17 people who

test positive were not in treatment during the 6 months before the test, only 2 were in treatment.

In conclusion we believe that if the first aim of methadone treatment should be the discontinuance of heroin use, of equal importance should be the prevention of HIV infection and the acquisition of a more socially stable life-style. The most relevant objective should be to ensure that a diagnosis and treatment plan is arrived at via a multiprofessional team, who have specific technical and cultural experience within the field.

### *The Padova Centre Report*

**F. Schifano**
Servizio Tossicodipendenze, U.L.S.S.n.21
Ospedale dei Colli, Via dei Colli, 4. 35100 Padova, Italy

The Padova Local Health Unit is the largest centre for the number (approx. 380.000) of inhabitants in our country. According to our estimates, 1,750 people from this catchment area are opiate users; 820 of them have been recruited up-to-now by our Addiction Treatment Unit (ATU). About 500 people attend the Unit daily and 400 of them are prescribed with methadone.

As described by other Mediterranean countries, our Unit also has a high rate of infectious diseases: 38% of the known opiate users are HIV positive. From another point of view, we consider that the policy in our Unit (multimodal treatment approach; emphasis on harm reduction risk and on psychiatric comorbidity) is quite successful, since about 72% of the clients, at the 12 months follow-up are still in contact with the ATU.

The ATU has been a member of the ECCAS network and an active participant in the Methadone Study from the very beginning, and we can report that we have not

**Table 1.** Characteristics of the sample

| | Seropositive n=42 | Seronegative n=50 |
|---|---|---|
| Male | 76% | 70% |
| Mean Age at first drug use | 19.8 | 21.9 |
| Mean Age at first HIV test | 24.8 | 24.5 |
| Mean Age | 30.5 | 27.9 |

**Table 2.** Changes in sexual behaviour knowing HIV status

| | Seropositive n=42 | | Seronegative n=50 | |
|---|---|---|---|---|
| | Before | After | Before | After |
| Mean No of IV admin. per week | 20.7 | 12.4 | 16.8 | 10.5 |
| % of needle sharing IVDU | 86% | 36% | 70% | 36% |
| Mean group size of needle sharing | 3.3 | 2.0 | 2.1 | 1.5 |
| % of needle sharing | 24% | 6% | 12% | 6% |
| % homogeneous mixing | 52% | 62% | 18% | 21% |
| Mean No of partners/yr | 4.1 | 1.4 | 3.5 | 3.1 |
| Mean No of intercourse/partner | 37.1 | 56.3 | 30.6 | 31.1 |
| % IVDU partners | 40% | 46% | 32% | 24% |
| % use of condoms | 5% | 38% | 12% | 21% |

encountered too many difficulties in recruiting patients (given the large number of those attending the Unit and since methadone prescribing, in Italy, is subject only to clinical limitations: if the doctor wants to prescribe, he can so.)

On the whole some interesting results from our work have begun to emerge since joining the project. In particular, we have compared two groups of patients (one in which the HIV test result was positive, the HIV+ve group, and the other in which the test result was negative, the HIV-ve group), with respect to possible changes in risk behaviours. The HIV+ve (n=42) and the HIV-ve (n=50) groups were comparable in terms of sociodemographic characteristics, mean age at first drug of use and mean age of first HIV test. As it is seen from tables 1 and 2, however, there were differences between the two groups, with respect to the direction and amplitude of the behavioural changes. In the HIV +ve group, after being aware of the test result, a reduction was recorded in: the mean number of I.V. administrations per week, the rate of needle sharing and I.V. administrations per week and lastly the mean number of people sharing injecting equipment. On the other hand, the rate of homogeneous mixing ( a concept with which we have attempted to derive if one is used to share the needles with anybody, in a "homogeneous" way, or only with selected, more "reliable" people) remained substantially stable (from 52 to 62%).

With respect to the sexual behaviour, in the HIV+ve group, after being aware of the HIV status, a dramatic reduction in their number of partners per year was observed, together with a rise both in the condom use and in the mean number of sexual intercourses per year with the partner. The mean number of I.V. drug using partners remained stable.

In the HIV-ve group, as in the HIV+ve group, a reduction was observed in the drug I.V. uses per week and similar modifications in the other parameters describing the number and characteristics of needle sharing. However, a lower rate of homogeneous mixing was already recorded, with respect to the HIV+ve group, in the pre-test era. In other words, it seems that this parameter, in comparison to the others, better differentiates between the two groups in terms of risk behaviour. With respect to the sexual behaviour, in the HIV-ve group it was not observed substantial modification in the parameters considered; however, one should emphasize that, in this group, the condom use is not very high (12% at baseline, 21% after the test).

Although our results have to be taken with some caution (given the small size of the sample and the retrospective design of the study) one could conclude that the decision of doing the HIV test, in a drug addict, turns out to be, in itself, both a way to better define the clinical situation of a patient and a tool to positively modify his/her risk taking behaviours.

## A Report from Spain
## The Methadone Maintenance Programs in Barcelona

**M. Torrens, L. San**
CAS-Barceloneta. S. Psyiquiatria y Toxicomanias Hospital del Mar,
Universitat Autonoma Barcelona, Barcelona, Spain

The most important health care problems among addicts consuming drugs by the intravenous route are those associated with HIV infection. According to the data from the "Registro Nacional de Casos de SIDA" (National Register of AIDS Cases), until the end of 1992, 11,353 cases of AIDS were diagnosed among addicts consuming drugs by the intravenous route (1) . Spain is the European country with the largest number of drug-related cases of AIDS followed by Italy and France (2).

According to this information at the beginning of 1990, a substantial modification in the regulations governing opioid maintenance programs was established. The results were that it was easier for addicts to be admitted in methadone maintenance treatment, and there was an important increase in the number of patients on this treatment (1).

### *Characteristics of the Methadone Maintenance Program (MMP)*

The main aim of the MMP is the prevention of HIV infection. In this sense low threshold programs were implemented in order to engage as many patients as possible on treatment.

There are minimal entry criteria, for instance, previous treatments are not necessary and it is accepted that free request by the patient in the absence of medical contraindication can be an entry criteria.

There are some priorities: pregnancy, AIDS, HIV infection, tuberculosis, and other psychiatric disorder different from drug abuse or dependence.

There are also minimal forced discharge criteria, and they are related to violence (physical or verbal aggression to other patients or staff), disruptive behaviour in the centre (e.g. drug trafficking) and/or non-attendance. There is no discharge for positive urine screening tests. When the urine control is positive the sanction is the withdrawal of privileges within the program (e.g. take-home dose).

There is not a time limit for the MMP, and the decision of discharge from treatment is related to individual characteristics and circumstances of each case.

There is not an upper limit for the prescribing of methadone, and the mean dose used is 80mg/day with a range from 25 to 230 mg/day.

*Health Interventions*

There is always a clinical evaluation, including: physical examination, hemogram, biochemical profile, HIV antibodies, hepatitis B and C screening, VDRL and, tuberculin skin test.

A hepatitis B vaccination program is offered and when it is necessary, control of tuberculosis treatment is done. Methadone patients routinely receive HIV counselling and advice about harm reduction techniques.

## References

1. Plan Nacional sobre Drogas. Delegacion del Gobierno para el Plan Nacional Sobre Drogas. Informe de situacion y memoria de actividades, 1992 Minsiterio de Sanidad y Consumo, Madrid, 1993
2. Reisenger M. Aids and drug addiction in the European Community. European Monitoring Centre for Drugs and Drug Addiction (EMCDDA). Commission of the European Communities, 1993

# A Report from Switzerland
# Methadone Treatment in Switzerland

**M. Bourquin**
Fondation Phenix, 8 bis Avenue des Communes Reunies, 1212 Grand-Lancy, Geneva, Switzerland

Methadone has been used in Switzerland for the treatment of heroin. A selection of papers was presented at the above workshop. The purpose of the workshop was to introduce ECCAS, present the preliminary findings of its first major study and present short papers on some of the work and the experiences of the participating centres.

## *The ECCAS Network*

Established in April 1992, ECCAS (European Collaborating Centres in Addiction Studies) is a developing network of professionals committed to advancing current knowledge in the field of substance misuse, which should have an impact at all levels: clinical, research and policy. The Network currently consists of sixteen centres in fourteen European countries: twelve member states of the European Union, Poland and Switzerland.

Since its inception ECCAS has been co-ordinating a large European Study, in which thirteen of the sixteen centres are taking part , investigating methadone substitution therapy and its impact on HIV risk behaviours; the background and initial findings are presented later in this paper. Apart from co-ordinating the largest study of this nature ever undertaken in Europe, ECCAS is also involved in a variety of other activities. Although at an early stage in its development, ECCAS produces a bi-monthly newsletter, and has plans to expand its program to include education and training initiatives, staff exchange programs, to co-ordinate and facilitate conferences on all matters related to the addictions, and to continue with its collaborative research program. With respect to the latter, ECCAS has already been involved in developing several research protocols, including funding applications.

# 25
# 1st European Methadone Conference Opening Remarks

**M.W. Parrino**
American Methadone Treatment Association, Inc.

**Summary**

The American Methadone Treatment Association is working to educate the public about methadone maintenance treatment. It is hoped that the recently published State Methadone Treatment Guidelines will bring greater stability for the treatment system of the United States. We will continue to work in conjunction with researchers and policy officials to improve treatment techniques. We will also continue to work with our International colleagues so that opiate users will be able to access good quality methadone treatment whenever they need access to such care. The decade of the 1990s will be critical for methadone maintenance treatment as new medications are made available at the program site. The Association will continue to organize methadone treatment providers throughout the United States to ensure greater program stability in the uncertain future of National Health Care reform and limited fiscal resources.
I would like to conclude by thanking my colleagues for their gracious invitation, which allowed me to share the perspectives of the American Methadone Treatment Association with the Conferees in Pisa. We will continue to work with you in creating a better future for methadone maintained patients and their families.

I would like to begin this plenary presentation by thanking Professor Castrogiovanni from the University of Pisa for supporting this Forum. I also want to express my gratitude to three remarkable individuals, who have made this Conference a reality. Dr. Icro Maremmani in Pisa, Prof. Alessandro Tagliamonte in Siena and Roberto Nardini in Pietrasanta, who have worked together in unifying methadone treatment in Italy. This 1st European Conference took enormous commitment and great vision. I pledge the continued support of the American Methadone Treatment Association to continue working with our colleagues in Europe and throughout the world. It is the collective hope of the Association's Board of Directors that methadone maintenance treatment will be recognized throughout the world as one of the most effective treatments for opiate dependence.

We are also fortunate to have three extraordinary people in our audience, who will be making presentations. Dr. Vincent Dole, Professor Emeritus of Rockfeller University who co-founded methadone maintenance with Dr. Marie Nyswander. Dr. Mary Jeanne Kreek is also present and worked with Drs. Dole and Nyswander from the very beginning in researching the use of methadone in treating heroin addiction; We will also hear from Dr. Robert Newman, President of Beth Israel Medical Center, which continues to treat the largest number of methadone maintenance patients in the world.

A large number of people are in the audience, from the United States and European Countries, who have devoted themselves to the work of methadone treatment. I want to express my gratitude to them for their efforts in understanding opiate dependence and their interest in providing humane treatment.

Many forces are shaping methadone treatment programs in the United States. President Clinton is proposing a major reform in how health care is delivered to U.S. Citizens. The needs of patients are also rapidly changing due to the terrible realities of HIV spectrum disease and the increased incidence of Tuberculosis. Most U.S. Methadone treatment programs are caught in the terrible conflict of needing to provide their patients with ready access to more complex services with fewer resources. It is important to explore the early support for methadone maintenance treatment in the United States and to review some of the best policy-driven research, which has been performed, prior to discussing future policy and treatment initiatives.

**Early Support for Methadone Treatment in the United States**

Methadone treatment was developed during the 1960s against the backdrop of increasing heroin addiction and the social turbulence of the Vietnam War. Cases of contagious serum hepatitis were on the rise and emergency room heroin overdoses are common. The early political support for methadone treatment in the United States was rooted in the correlation between untreated heroin addiction and active criminal behavior.

A retrospective study of methadone treatment admission and drug-related arrests were conducted in the five boroughs of New York City by Drs. Dole, Dejarlais and Herman Joseph. They found that approximately 20.000 heroin addicts were admitted into methadone maintenance treatment between January 1, 1971 and December 31, 1972. Drug-related arrests decreased by 24.000 and there was a reported decrease in burglary complaints and robberies by 77.000 during the same two year period. Reported cases of serum hepatitis also decreased by 1.500. Methadone treatment programs had their greatest expansion in the United States during the last 1960s and early 1970s.

The U.S. Food and Drug Administration promulgated comprehensive treatment requirements in 1974 concerning patient admission criteria, methadone dosage requirements for initial and maximum dosages and a series of staffing and treatment requirements. Individuals states were also in the process of promulgating their own methadone treatment regulations. At times, the state requirements were even more stringent than the federal requirements.

The initial political and public support for methadone maintenance treatment began to fade in the late 1970s and continued through the 1980s. Communities objected to having methadone treatment programs sited in residential settings. A number of dysfunctional and highly visible patients would loiter in the vicinity of the Clinic, intensifying the community's negative reaction. Many treatment programs closed during the decade of the 1980s and few opened. As a result, a larger number of patients were treated in fewer methadone treatment programs, this was especially true in large cities such as New York.

The pendulum of political support for methadone treatment swung back in favor of such interventions when HIV infection rates increased dramatically in the IV drug-using population. Methadone treatment programs were seen as a means of containing the spread of HIV Spectrum disease and AIDS, hence, protecting the public's health and well being. The U.S. Congress was at the point of implementing funding initiatives to support an expanded treatment system in confronting the needs of patients in the current era.

Congressman Charles Rangel, Chairman of the Congressional Select Committee on Narcotic Abuse and Control, directed the General Accounting Office to conduct a critical evaluation of U.S. Methadone treatment programs. The GAO Report concluded "that methadone is the most effective method available for treating heroin addiction... In practice, however, the continued use of heroin that the General Accounting Office found among patients in 24 methadone maintenance treatment programs indicated that many programs are not effectively treating heroin addiction... The use of heroin by patients in treatment for more than 6 months ranged from 1% at one program to 47% at two others." This study was also significant in finding that the policies and practices of the 24 methadone programs under study varied significantly. This GAO Report recommended the development of objective outcome-oriented criteria in evaluating the long-term success of methadone treatment. It also recommended a more sustained federal oversight in monitoring the work and effectiveness of methadone treatment programs. Ultimately, the congressional Committee was dissolved two years after the study was published. Unfortunately, Congressional Representatives do not have a favorable view of methadone maintenance treatment. A significant portion of the American Methadone Treatment Association's work is to educate Congressional Committee about the value of methadone treatment.

## Critical Policy-Driven Research in Support of Methadone Treatment

The methadone treatment programs of the 1970s were treating patients with comparatively less complex illnesses, which responded to less costly interventions. The methadone treatment programs of the 1990s are encountering patients with increasing morbidity, driven by HIV Spectrum disease and increasing multiple substance abuse. The National Institute of Drug Abuse has supported an enormous number of research initiatives in evaluating methadone treatment programs. The majority of these studies evaluated the biomedical effects of addiction and focused on the individual patient. Most of these research was not focused on changing policy. Methadone maintenance treatment research has become more action-oriented as a result of the work by Dr. John Ball and Dr. Thomas McLellan and their respective associates. Dr. Ball and his colleagues conducted a three year study through a NIDA Grant, which evaluated six methadone treatment programs in Baltimore, Philadelphia and New York. Dr. Ball's study arrived at a powerful conclusion: "Marked differences in the effectiveness of various programs were observed: Current IV use varied from less than 10% to over 57% of patients in particular programs". Dr. Ball also determined that the most effective treatment programs had high patient retention rates and a good relationship between caring staff and patients. The least effective programs prescribed inadequate methadone doses and had high rates of staff turnover. Patients characteristics were similar for all programs under study, therefore, the treatment environment emerged as a powerful indicator of patient success.

The importance of Dr. Ball's findings cannot be overemphasized. Policymakers must be aware of the elements, which contribute to effective treatment interventions. More than thirty years of research have indicated that retaining the patient in methadone treatment improves their health and creates stability. Fifty-one percent of the patients in Dr. Ball's study remained in their respective methadone treatment programs for a two year period. The remaining forty-nine percent remained in treatment for three to eighteen years. Conversely, 82% of the patients, who either elected to discontinue treatment or were discharged from treatment, relapsed to heroin use within twelve months of taking their last prescribed methadone dosage.

**The Development of the State Methadone Maintenance Treatment Guidelines**

The GAO Report and Dr. Ball's study indicated that the great variability in the quality of treatment among the nation's 750 methadone treatment programs was having a negative impact on patient care. These methadone treatment programs operate in forty-one states, treating approximately 115.000 patients on any given day. The American Methadone Treatment Association, in conjunction with the American Society of Addiction Medicine's Committee on Methadone Treatment, developed practical treatment guidelines for methadone programs under the sponsorship of the Center for Substance Abuse Treatment, which is part of the U.S. Department of Health and Human Services. The Guidelines were written to serve as a model for state policy officials and methadone providers in using the most effective therapeutic techniques in support of good quality patient care. These Guidelines were widely distributed in November, 1992 during a National Methadone Conference. The two prevailing themes, which were expressed in each chapter of these Guidelines included the need to respond to individual patient needs and offer comprehensive treatment services at the program sites. The chapters and authors for these Guideline were chosen with great care, responding to the most troubling issues, which were negatively impacting on patient outcome. The chapter on methadone dosage discussed the rationale for prescribing therapeutic methadone dosages for the patient. Chapters were written to guide the clinician and policymaker alike on the most effective methods of treating multiple substance abuse, the most effective urinalysis testing techniques, in addiction to reviewing admission criteria and effective methods of treating pregnant methadone maintained patients.

These Guidelines also presented the most critical elements in constructing comprehensive patient profiles so that a treatment program could respond to the specific needs of their patients. A methadone program manager could work in con unction with treatment personnel in developing a comprehensive treatment profile, which could be updated each year to incorporate the following elements of patient care. The program manager/coordinator would then be able to match the appropriate mix of treatment personnel to the needs of the patient, determining the best range of services to be provided with the available funds. The following seven elements were recommended as primary in developing this comprehensive patient profile:

1. The number of suspected or known AIDS cases
2. The number of estimated or known HIV-positive patients or patients infected with TBC or STDs
3. The trends in alcohol, opiate, cocaine, and other forms of substance abuse, including routes or administration
4. The number of vocational counseling and job referral candidates
5. The number of patients in need of specialized medical care
6. The number of pregnant women
7. A comprehensive morbidity index for all patients in treatment.

**Structuring Methadone Treatment Programs**

Each treatment program operating within a particular country's system of care must be designed in such a way so that the patient improves as a result of entering and remaining in methadone treatment. Policies should encourage the patient to access treatment as needed, meeting the appropriate clinical requirements of the treatment program. The patients's methadone dosage should be adequate so that the patient does not need to use heroin or other substances. The patient should be treated respectfully in a therapeutic environment and should continue in treatment as long as the patient benefits from ongoing care. The decision to continue in treatment is best made between the patient and the treatment practitioner, free from outside intervention from government and other external authorities.

Each country's treatment practitioners can draw upon the extensive methadone treatment research that exists in addition to more than thirty years of clinical experience. Treatment practitioners, in conjunction with government and funding authorities, must determine what kind of intervention needs to be provided in order to effectively treat opiate dependency. If HIV Spectrum disease is significant in a particular country's population, then treatment caseloads and interventions should be configured to provide better access to HIV counseling in addition to intensive clinical support for treating alcohol and drug using patients. Positive and sustained outcomes can only be achieved in a stable therapeutic environment, which is staffed by well trained and supportive caregivers.

**Methadone Treatment Policy Currents in the United States**

There are two competing currents among policymakers in the United States concerning the expansion of methadone maintenance treatment. The first group believes that treatment programs must be responsive to changing patients characteristics by providing more services at the program site and providing adequate medical care to treat increasing patient morbidity. The second group insists that government agencies cannot sustain such a treatment intervention and believes that patients can be treated with limited resources. Dr. Thomas McLellan and his colleagues conducted research, which evaluated the impact of providing different levels of care in methadone treatment. Dr. McLellan's research findings clearly support the use of comprehensive treatment interventions in responding to individual patient needs, if that is what patient requires. These findings are consistent with other research, indicating "That those substance abuse patients who receive the most services during treatment have the best outcomes; and that those substance abuse treatment programs that provide the most services to their clients have the best programmatic results".

Dr. McLellan's study evaluated three levels of methadone treatment services, which were monitored over a six months period. The patients in each of the three trials received a daily dose of 60 mgs. of methadone. The Minimum Methadone Service Trial was limited to dosage administration and crisis-oriented counseling services. The Standard Methadone Service Trial provided regular counseling services in addition to daily dosage administration. The Enhanced Methadone Service Trial provided the richest compliment of services, including regular counseling, medical and psychiatric care, family therapy and employment counseling. The best outcome was observed with the patients who were involved in the Enhanced Methadone Service Trial. These patients showed the most dramatic decrease in heroin use of all the research trials, even though patient characteristics were similar throughout and the methadone dosage was also similar throughout.

# 26

# Incorporating a Methadone Maintenance Program into a Drug-free Treatment System

**M. A. Rizzi**
CODAC, Cranston, Rhode Island, U.S.A.

## Summary

The intent of this presentation is to identify and address obstacles to the establishment of a methadone maintenance treatment program, not only within a community or neighborhood, but within the operating structure of a drug-free, abstinence-based, outpatient treatment program. Opposition to the development of methadone maintenance treatment is not necessarily limited to community and political constituencies. Often, treatment staff, steeped in the tradition of a drug-free, or abstention-based, model of recovery may be an equally adverse sub-group. Treatment staff that is not convinced of the efficacy of methadone maintenance treatment and/or is not able to objectively consider its proper application, may not only negatively influence the acceptance of methadone maintenance treatment service but outcome as well. Likewise, treatment staff with a history of providing services oriented towards abstinence is faced with a methodological and philosophical dilemma. They must address questions regarding their own perceptions and beliefs. Program personnel must be educated, their anxieties addressed and they must be given support as they are faced with the questions related to a radically different approach to the treatment of opioid dependence. Additionally, drug-free, or abstinence-based programs, must contend with the attitudes and perceptions of those patients, currently in treatment and following an abstinence model. They also will question the use of methadone or any other pharmacotherapy. Drug-free patients will require the same type of education but they also must be encouraged to understand that pharmacotherapy for others neither diminishes nor undermines their own recovery. This presentation will discuss how CODAC, a non-profit substance abuse treatment program, with 18 years of methadone maintenance treatment experience, was able to overcome community, staff and patients resistance in the development of an effective methadone maintenance program in Newport, Rhode Island, U.S.A.

The intent of this presentation is to identify and address some of the essential obstacles to the establishment of a methadone maintenance treatment program, not only within a community or neighborhood, but within the operating structure of a drug-free, abstinence-based, outpatient treatment program.

It should be clear that a methadone maintenance treatment program, or any other human service provider, is a member of the community - it can not stand alone, in defiance or in arrogance. As a member of the community, it is a neighbor - a participant in the health and well-being of that neighborhood. Those responsible for the management of a clinic must be ambassadors as well as clinicians, politicians as well as administrators. They must make their clinic a good neighbor.

It is of considerable concern that many communities in the U.S.A. have been opposed to the siting of methadone maintenance treatment programs in their neighborhoods. This opposition has been based upon a number of factors:

1. general opposition to any mental health or human services organization - they believe that social services are important but "Not in my backyard", otherwise known as "NIMBY".
2. specific opposition to substance abuse treatment
3. specific opposition to methadone maintenance treatment programs.

This opposition is generally motivated by concerns regarding their:

1. animosity, aversion, apprehension, and anxiety related to drug addicts in general
2. lack of tolerance for drug abusers, specifically opioid users
3. concerns regarding reduced property values
4. community image
5. increased drug traffic and paraphernalia on local streets
6. drug related crime, i.e., theft, larceny, assault, rape, etc.
7. HIV/AIDS-related risk
8. their perception about methadone maintenance treatment.

Opposition to the development of methadone maintenance treatment is not necessarily limited to community and political constituencies. Often, treatment staff, steeped in the tradition of a drug-free or abstention-based model of recovery may be an equally adverse sub-group. Treatment staff that is not convinced of the efficacy of methadone maintenance treatment and/or is not able to objectively consider its proper application, not only influences the acceptance of methadone maintenance treatment service but outcome as well. Treatment staff is also faced with a methodological and philosophical dilemma. They must address these questions:

1. Does methadone maintenance treatment challenge their concept of recovery?
2. Can they support replacement drug therapy for some and abstinence for others?
3. Does methadone maintenance treatment imply that abstinence is unacceptable?
4. Does methadone maintenance treatment challenge the effectiveness of abstinence?
5. Can methadone maintenance treatment co-exist with abstinence?
6. Can methadone maintenance treatment patients be considered in recovery?
7. What is recovery?
8. How can they justify methadone maintenance treatment to a drug-free patient who is struggling with his or her own recovery?

Additionally, drug-free, or abstinence-based programs must contend with the attitudes and perceptions of those patients, currently in treatment and following an abstinence model. They too, have anxieties regarding their own recovery, have heard so much about what it means to be drug-free, have been led to believe that true recovery is the total freedom from any dependence and that methadone is just another drug, another crutch that prolongs addiction.

To develop a methadone maintenance treatment program, the needs, fears, anxieties and apprehensions of many must be addressed. Neighbors must be heard, educated and assured, as much as possible, that neither they nor their property is at risk. Health care officials must begin to understand that methadone maintenance treatment program not only eliminates behaviors associated with addiction, is less expensive than other forms of treatment, is an excellent tool in the prevention of HIV-related disease and is certainly less costly than the medical care associated with the medical treatment of HIV.

Likewise, program personnel must be educated, their anxieties addressed and they must be given support as they are faced with the questions related to a radically different approach to the treatment of opioid dependence.

Lastly, drug-free patients will require the same type of education but they also must be encouraged to understand that pharmacotherapy for others does not challenge or undermine their own recovery.

There has been a long-standing history and tolerance of alcohol use in this resort, navy town. However, like other resort communities, illicit drug use and traffic, an increasing transient population looking to party and have a good time, have created an environment in which substance-related problems grow quite freely. It is said that substance abuse flourishes when there is opportunity, demand, supply and poverty. Newport, unfortunately has all of the above.

In 1987, the School of Public Health at prestigious Harvard University issued a report entitled "A Drug Abuse Treatment and Prevention Plan for Rhode Island". This report was commissioned by the state's Department of Mental Health, Retardation and Hospitals for its Division of Substance Abuse. This plan made recommendations in many areas, but specifically addressed the unmet need of IVDUs (intravenous drug users). Their recommendations included the following:

1. HIV-related services for IVDUs needed to be expanded
2. demand for service for IVDUs, specifically opioid users, far exceeded available treatment capability
3. methadone maintenance should be seen as a responsible tool in the effective management of the treatment and spread of HIV-related disease
4. methadone maintenance services needed to be expanded and made more accessible
5. methadone maintenance services should be provided in the greater-Newport area.

For all the reasons listed above, the recommendation to establish a methadone maintenance treatment center in Newport was given serious consideration by state regulatory bodies.

The Rhode Island substance abuse provider community has been fortunate to have an enlightened and supportive regulatory and legislative infra-structure. Those elements that fund and oversee substance abuse treatment have long recognized the need for a comprehensive and inclusive system of treatment and prevention services. Consequently, four (4) state regulatory bodies - Department of Health; Division of Drug Control; Department of Mental Health, Retardation and Hospitals; and the Division of Substance Abuse along with two (2) federal regulators - Food and Drug Administration (FDA) and the Drug Enforcement Administration (DEA) worked hand in hand to facilitate the expansion of methadone services in Newport.

CODAC, Inc., a private, non-profit organization with twenty (20) years as a provider of out-patient, drug-free substance abuse services and 15 years as a methadone maintenance provider was selected as the provider for these services in Newport. CODAC had also operated a small, drug-free clinic in the heart of Newport's business district since 1978.

While regulatory support and organizational support was high, once these intentions were made public, the three (3) constituencies mentioned above - neighbors, staff and patients - were quick to voice objection. Each group had to be won separately. It became clear that a well-coordinated effort, combining the marketing skills of the regulators, CODAC administration, elected officials, a motivated landlord and the press would be necessary to actually implement the program.

While the community endorsed and supported the need for the "traditional" drug-free services that had been provided, and would continue to do so, they would not go so far as to condone a methadone clinic.

Concurrently, and to our surprise, a small, but growing support network was building. Elected officials, the press, families of recovering addicts, former and current methadone patients, other concerned citizens, an alternate landlord and law enforcement officials banded together to voice their opinions as to why there should be a methadone clinic in Newport. It is important to note that the press not only reported this controversy objectively, it also authored editorials very much in favor of methadone treatment and very much against the unreasonable attitudes and behaviors of local citizenry. Their support eventually resulted in the approval and siting of the clinic, drug-free and methadone at another, nearby, location.

It was, however, important to listen to the concerns of our prospective neighbors and to respond respectfully to them. To minimize their worries or to not give them due consideration would have only escalated their objections and resistance. A combination of facts in the form of data and literature, most specifically the writings of John Ball and Mary Jeanne Kreek, coupled with concrete responses to some of their concerns, served to soften their objections to some degree. By limiting dosing hours so as not to create parking problems or to interfere with normal commerce in the area and by assuring that loitering around the clinic would not be tolerated, our neighbors agreed to withhold their protest until they had cause.

Initiating methadone services within our existing drug-free clinic was another challenge.

In much of America, methadone treatment is seen as the end of the treatment continuum. This perception was not too different in our own clinic. Even though CODAC had conducted methadone services for over 15 years, the staff and management of the Newport clinic was not convinced that this form of treatment was anything more than a "legal" addiction, and not at all recovery. It would have been relatively easy to enact by ultimatum but it is clear that the resulting negativity would have been overwhelmingly destructive. Staff needed to be re-assured that methadone maintenance treatment did not diminish the importance of their drug-free work. Some felt that the onset of methadone maintenance treatment was a refutation and minimization of their earlier work. In anticipation of this resistance, supervisory staff was invited to participate in clinical and administrative meetings held at our methadone maintenance treatment program in Providence. They were encouraged to question any and all existing policies and procedures in order to get a sense of the philosophy and direction of that program.

They were also encouraged to interact with patients and prospective patients in similar fashion. Journals and other literature dedicated to methadone treatment and research was provided to enhance their understanding of the history and progress of methadone maintenance treatment. Additionally, studies that addressed the relationship between methadone maintenance treatment and HIV- related disease were made available.

Methadone maintenance was shown to be more than a medication program, but an integrated system of medical, clinical and social services designed to provide the patient with a comprehensive and compassionate response to the chronic, relapsing disease of opioid dependence.

As supervisory staff became more comfortable and educated, they were encouraged to share their new-found information and perspectives with their subordinates. Concurrently, they began to prepare their staff for the eventual incorporation of methadone maintenance treatment at the clinic. It was immediately clear to all that a well-run methadone clinic is highly structured, has a specific limit-setting orientation and requires much more patient compliance than a typical drug-free clinic. Many of the freedoms allowed on the drug-free side were contraindicated in methadone maintenance treatment.

Some staff saw the type of limit-setting associated with methadone maintenance treatment as punitive rather than constructive. For instance, a positive urine for a drug-free patient might result in a discussion of the events regarding the unauthorized drug use while the same positive urine specimen for a methadone maintenance treatment patient would result in the discussion, a contingency contract and the loss of take-home privileges. The inconsistencies in patient management techniques was seen as confusing and alienating. They were accustomed to being seen as "good guys" - now they would be required to engage in policy implementation and run the risk of alienating their patients.

A considerable amount of time was spent dealing with this issue from a both practical and subjective perspective. Some were uncomfortable with the "power" that seemed to be inherent in their new responsibilities while others needed to curb their inclinations to exercise that "power".

Medical, nursing and clinical staff from the Providence program provided inservice training for the Newport staff and the Newport staff, like their supervisors, was advised and encouraged to visit not only the Providence clinic, but other methadone maintenance treatment programs in nearby Massachusetts and Connecticut. During this process it was vitally important that the "educators" remain sensitive to the needs, concerns, and apprehensions of the clinical staff. Questions and issues were processed so that all understood the intent of the question and the implication of the answer. The policies and procedures manual of our existing methadone maintenance treatment program was reviewed and discussed in its entirety. The background for each policy was examined and the previously considered alternatives discussed. Answers resulted in further questioning.

The process continued until all felt reasonably comfortable and able to translate this new information into practice.

You must remember that the first methadone maintenance patient had not even entered treatment at this point. The task now before program staff was to prepare for incoming patients and to deal with any issues raised by current drug-free patients.

This task proved to be easier than expected and was probably the result of effort made by staff to understand methadone treatment but to convey this information to their patients in the form of individual and group discussions. The educational and informational needs were handled in much the same way as the previously referenced staff orientation and education. Concerns and apprehensions were addressed.

While methadone dosing services were provided during limited hours in order to satisfy the neighborhood need for traffic control, clinical services were integrated into the day-to-day activities of the existing drug-free clinic. The result was a therapeutically diverse co-mingling of drug-free and methadone patients in a variety of activities. Drug-free and methadone patients participate together in group counseling sessions focused on specific issues, i.e., problems with alcohol and/or cocaine, parenting, grief and loss, Narcotics Anonymous, etc. The result of this interaction is a greater tolerance for, and understanding of, the treatment and support needs of others. In some instances, methadone maintenance patients have been encouraged to consider withdrawal and abstinence with a support group heretofore unknown. What all have realized is that each patient is striving for the same thing - the opportunity to live a life that is free from the stress, anxiety and risk associated to drug dependence. Each is seeking recovery that enables him or her to re-engage, or perhaps engage for the first time, with the mainstream of society and to become a productive, responsible and happy human being.

We are pleased to report that in the four years since this program began, we have been recognized as a good neighbor and have been responsive to our community. Those who at first objected, including our staff, have come forward and informed us that their opposition was premature. The local press has written a status report which stated that initial fears and concerns were unfounded. The chief of police has stated that crime in the area of the clinic has decreased since the clinic began operation.

Perhaps we have been lucky. Perhaps, good planning and the ability to respond rather than react helped turn the tide. We believe that it is a combination of many factors in addition to the above - dedicated supervisors and staff, support in high places, an objective press, politicians willing to take risks, and for the most part, a community willing to address its problems.

# 27

# Heroin Dependence Theory of Different Levels of Intervention

**I. Maremmani[1], O. Zolesi[1,2], L. Daini[1,2], R. Nardini[3], P. Castrogiovanni[1,2]**
[1]Dependence-Psychiatry Unit , Psychiatric Institute, University of Pisa, Italy
[2]Ph.D. Research Program Siena, Pisa & Cagliari Universities
[3]Social Disease Study and Intervention Association

**Summary**

The medical, psychiatric, psychological and social manifestations of heroin addiction require more than a multidisciplinary intervention. To effectively treat addiction, rehabilitation and/or prevention is necessary but phase of illness must also be treated. In fact, contrary to what is usually believed, therapeutic interventions for drug addiction (rehabilitative drug-free treatments and chemotherapeutic interventions) are not singularly sufficient nor are they interchangeable in the single subject. It is often necessary to adapt the intervention to the clinical phase of illness, trying to enhance program "retention rate". This condition is indispensable in the rehabilitative process. The nature of drug addiction will often necessitate contacting clients in the street, in order to begin counselling and "harm reduction". Finally, primary or secondary prevention cannot be separated from a global intervention philosophy.

Given the complex nature of heroin dependence there is no one method that is completely effective in the treatment of this pathology. Drug addiction varies in intensity and drug addicts are a heterogeneous group in terms of personal resources and coping ability.

The clinician's priority, as recent Italian legislation states, is to respond appropriately to each individual patient, by personalizing therapeutic planning (including different types of interventions) creating an articulate form of treatment in an effort to improve the single drug addict's functioning. Each single treatment should be interchangeable and adaptable to the patient's changing needs thus providing long term continuity.

Presently, almost all researchers, in the field of drug addiction, are in agreement that the "retention rate" is a fundamental requisite for the successful outcome of any program (1). This is obvious if one considers the official definition of drug addiction as a chronic and relapsing illness. Thus therapeutic planning must be adapted to the patient and not vice versa (2).

If, as systematic observations reveal, many drug addicts may remain such for a long time, some for the rest of their lives, attempts to treat this vast group of subjects must not be abandoned. It would be sufficient to opt for long term treatment giving the drug addict the possibility to gradually recuperate bio-psycho-social functioning. This could be defined as clinical recovery even if "restitutio ad integrum" has not been achieved and is the first goal to be aimed at with adequate pharmacotherapy and cognitive behaviourally oriented treatment (3-6).

Achieving this goal may be the best results possible for some, while for others it may open the door to a drug free state often unobtainable if proposed to individuals who are not yet rehabilitated and reinforced in their psychological and social functioning. In both cases however, these subjects have a right to a normal life, to personal gratification, social respectability and physical and mental well-being.

The second step in therapeutic planning is monitoring patients during and after treatment in order to prevent and treat relapses which, although undesired, are however, frequent. With this new viewpoint, relapses become predictable, expected; the operator and patient must be ready to face the occurrence with all the available resources in order to maintain the functioning level previously reached. Thus when a relapse occurs this should not be interpreted as a sign of failure, as often happens, nor should the patient be penalized or treatment denied as often is the case in Italy and in many European countries.

It is important that operators acquire a global view of the various types of treatment available, the length of time required, the indications and contraindications, as well as the possibility of cross over in order to obtain maximum therapeutic results for that single patient in that given moment (7).

If, on the one hand comprehensive treatment programs attempt to deal with many of the problems associated with addiction, we feel that treatment modalities based on segregation and accusation must be seriously revised. These methods are highly selective and beneficial to a very limited number of addicts. They may, sometimes, cause cognitive disorganization of the patient; they tend to reinforce guilt and give the idea that drug addiction is an acquired vice caused by deviant behaviour. In this way the patient finds himself defenseless and unprepared for relapses interpreting these to be an explicit sign of being irrecoverable.

Family counselling in these programs is very harsh and often implies cutting ties with the patient who is described as "lacking in will and motivation". "Reaching the bottom" for the drug addict very often means dying of overdose, or contracting AIDS or refusing all types of treatment.

Comprehensive treatments need a new generation of operators. They must knowledge of the various levels of the treatment program and that these levels are communicating.

Our theory of comprehensive treatment includes different levels of intervention which are: Level 0 : prevention. Level 1 : harm reduction. Level 2 : diagnosis and treatment of associated pathologies. Level 3 : specific treatments. Level 4 : rehabilitation and social integration. Level 5 : prevention and treatment of relapses.

**Level 0. Prevention**

Currently an efficacious primary prevention model does not exist. Operators propose educational models based on their own cultural background which are rarely acceptable to all.

Although drug dependence may have its roots in societal organization, in consumerism, educational models alone are not effective in the prevention against drug addiction and may cause diametrically opposite results in social groups with different cultural backgrounds (8) (9). Research has not discovered specific educational impairments, nor temperamental types associated with drug addiction. A large number of subjects begin using drugs for fun or to facilitate socialization without knowledge of the real consequences of drug abuse (10).

As a primary prevention model, we suggest a public health education program on the various psychoactive substances of abuse, effects, consequences of use and abuse, disregarding ideological and moralistic interpretations which often succeed in leaving a mythical and mysterious imagine which fascinates (11) (12).

If health education constitutes a valid primary prevention policy, secondary prevention (harm reduction, therapy, prevention and treatment of relapses) must not be overlooked.

Research indicates that the spread of heroin use correlates to precise market interests which are kept alive in certain well-defined conditions such as clandestineness which implies high cost, consumer-pusher phenomena and when effective therapy is lacking. Within this framework depenalizing drug use, controlled administration of heroin and treatment of addiction are three essential cornerstones in the elimination of this problem (13).

**Level 1. Harm Reduction**

The aims of level 1 may be summarized as follows:

a) reduce the social consequences related to addiction, such as: criminal activity, spread of AIDS, extinction of the consumer-pusher phenomena, elimination of the clandestine market with subsequent reduction of number of heroin users, and minor risks for the general population.
b) protect heroin addicts from syringe related pathologies (HIV, hepatitis B, vascular damage, endocarditis, overdose, withdrawal syndrome, etc.); this will prove advantageous for the patient and will reduce social costs.
c) more accessible public health services for the heroin addict population (14). Establishing the first contact between health operators and addicts means (1) reaching a larger number of subjects; (2) offering accurate information regarding physical and mental well-being and therapeutic prospects. These measures are lacking especially in Italy and in other parts of Europe.
d) the possibility of an early diagnosis which is presently impossible since drug addicts live in clandestineness. The patient usually seeks help when the situation is no longer bearable and course progression is well advanced.

Proposable interventions at level 1 include:

a controlled distribution of heroin and other agonist opioid drugs

a controlled distribution of heroin and other agonist opioid drugs
b free distribution of disposable syringes
c instructions regarding self-administration of medications.
d information regarding first aid in case of overdose or withdrawal syndrome
e information regarding the risks and consequences of continued use of illicit drugs and modalities of treatment and rehabilitation.
f health education of HIV subjects.

The operative phase of level 1 would be carried out by volunteers and specialized operators in "street units". Family physicians as well as ambulance paramedical personnel should also be involved. In this way a tight network of contacts between health services and drug addicts is assured and access to health services is facilitated.

The effectiveness of an antiprohibitionary and pragmatic approach is widely demonstrated in the experience of countries such as England and Holland which have succeeded in limiting the spread of heroin addiction (e.g. in 1991 in the United Kingdom 8.000 heroin addicts were officially registered; there were no deaths due to overdose, the spread of AIDS was limited and restricted to subjects at risk, prevalently homosexuals). In Italy with its moralistic and repressive attitude, in the same period there were more than 320.000 heroin addicts, 1,200 deaths by overdose, widespread diffusion of HIV and 70% of heroin addicts were seropositive.

The drawback of this first level is that it is not an actual treatment modality and therefore it cannot help patients recuperate bio-psycho-social functioning. In order to achieve this goal we must pass to the next level of our program which includes services and more qualified personnel.

### Level 2. Diagnosis and Treatment of Associated Pathologies

At this level the specific treatment of drug addiction begins. The patient is examined by a medical specialist and other professional personnel in order to arrive at a diagnosis and establish a therapeutic plan appropriate for that subject. Scientific literature is in agreement in defining heroin addiction as an illness and experience shows that it is the patient's degree of impairment together with other factors that determine if a particular intervention is suitable or unsuitable at that time. The principal task of the specialized operator at this stage is to formulate a diagnosis and identify potential resources (personal attributes, family members or social skills), that may help in rehabilitating the patient. This will be possible if interviewing techniques reactivate a two way communication in order to identify the needs of the patient and offer concrete proposals (15). Particular attention should be given to unsuccessful endeavours which are often indicative of errors in the interventions proposed or in monitoring of the patient.

This level requires more qualified personnel and specialized services. Specialist centres for the diagnosis and treatment of addiction are needed. These centres should be equipped to carry out research, collaborating with Ph. D. Research Programs in Drug Addiction, and educate operators and train specialized personnel.

Once a diagnosis has been made, the patient undergoes the appropriate therapeutic modality. The initial choices, however, should not be restrictive or rigid but rather open and interchangeable with other modalities. Only if the patient acquires and maintains a functional state will the operator be able to verify the choices made.

At the same time associated pathologies and psychiatric disorders are diagnosed and treated (16).

## Level 3. Specific Treatments

This level includes therapeutic and rehabilitate interventions after the patient has undergone clinical assessment. Generally patients may be divided into two groups:
a patients who do not require medications or opioid agonists.
b patients who require methadone maintenance.

### *A. Patients Who Do not Require Medications or Opioid Agonists*

The patients included in the first group (drug free or with opioid antagonist therapy) should satisfy the following requisites: they are subjects who **abuse** psychoactive substances according to the definition of DSM III-R; they have low craving; good social adjustment; good family support with the possibility of a referring family member; these subjects are reliable and have good interpersonal relationships with staff.

However it is important to underline that methods based on a "drug free state" are highly selective and applicable to a very small number of patients. This is probably due to the fact that drug addicts seek help in late stages of illness. If at this point diagnostic assessment is erroneous the patient may recur to heroin use concealing it from the operator. It is understandable, then, the caution needed before detoxifying patients, as well as, the need to control attentively behaviour at risk and immediately admit the patient to a methadone treatment program if difficulties arise.

Methods for achieving a drug free state may be outlined as follows:
1. Abstinence is controlled by psychotherapeutic support, with or without opioid antagonists.
2. Open Community Program. Self-help groups which encourage social reintegration during treatment. Antagonists may also be used in this case.
3. Residential Community. We suggest a revision of therapeutic communities programs in light of biological theories of drug addiction.

### *B. Patients Who Require Methadone Maintenance*

This group includes the large majority of drug addicts who seek help at public health services. They do not meet requisites for "drug free" programs which would be detrimental for these subjects. Nevertheless, these patients often insist on rapid detoxification from methadone. This request is supported by the erroneous notion that drug dependency is "headstuff" and related to will-power and thus a vice and not an illness. These notions are often reinforced and confirmed by poorly informed operators who consider drug addiction a guilt to expiate and not a pathology to treat.

The first task the operator must face is that of redefining the patient's expectations suggesting long term treatment which will probably be more successful and safer.

Let us look at services that may support and be integrated with pharmacotherapy (Methadone Maintenance)

1. Basic counselling. Many patients on methadone who have obtained metabolic stabilization experience a return to normality; they become socially reintegreted especially if they have personal resources, help from family members (home, work, hobbies etc). For these patients therapeutic success may be possible with specific information and treatment counselling.
2. Individual psychotherapy and/or treatment of psychiatric disorders for subjects with specific problems.
3. Open Community Program. Self -help groups could provide solid support to those subjects who lack rehabilitative resources. In future we suggest that more attention be focused on these groups because they are at low cost, have been shown to be effective in other areas (alcohol, psychiatric pathologies, etc) and more subjects can be treated simultaneously.
4. Residential Communities. These communities would serve those subjects who need specialized social structures in addition to pharmacotherapy. They are drug addicts with serious psychiatric disorders as well as those addicts who find themselves jobless and homeless.

In closing this chapter on treatment modalities we would like to underline:

1. The therapeutic communities we have outlined are quite different then those present in Italy today. They would be linked to social agencies and other health services. Unlike the existing communities they would no longer be reclusive structures and isolate the patient from his family and social ties. At present these communities create an artificial world in which recovery is obtained and quickly lost when the patient is released.
2. The primacy of "drug free" programs should be abolished. Recovery cannot be associated with a "drug free" state. It should be related to the psychological and social functioning.

**Level 4. Rehabilitation and Social Integration**

This level foresees the complete rehabilitation of drug addicts independently of the kinds of treatment modalities in progress. The achievement of this goal varies (length of time and modality) according to the needs and the severity of illness of each individual. The interventions which allow the patient to achieve this status vary, for example: getting a job, reintegration into family life; methadone detoxification.

We would like to focus the need of those patients who cannot be deprived of methadone due to consistent metabolic impairment. A substantial part of the drug addict population who have good social and psychological adjustment require methadone maintenance but not social support services. We consider these patients completely recuperated and feel that they are able to manage their pharmacotherapy i.e. as diabetics do. For these subjects methadone availability should be convenient and interfere as little as possible with the patient's life, work and leisure time. The patient could be entrusted with dosages that cover a longer period of time; family doctors would be able to prescribe methadone. Any community health service could dispense of methadone under certification in order to facilitate the patient. On international level contacts could be established between the health services of different countries permitting the patient to travel freely. The organization of a health service network would prove advantageous for the patient who need not travel great distances to reach specialized centres and at the same time these centres would not be overloaded with work-dispensing of methadone to patients who are rehabilitated, thus reducing social costs.

## Level 5. Prevention and Treatment of Relapses

Given the definition of heroin dependence as a chronic and relapsing illness, it is logical to emphasize the role of prevention and therapy of relapses. This requires therapeutic modalities which help in conserving the skills and functioning level previously achieved by the patient. Thus patients would be rapidly readmitted to methadone maintenance therapy (methadone is obligatory with relapses) in order to prevent harm to the patient i.e. returning to street life. Treatment would be simplified in these programs as these patients have been rehabilitated in the past. In order to accelerate readmission to any health service the patient would be provided with documentation containing clinical chart data.

## Conclusions

In conclusion, we have attempted to outline a rather complex strategy for the treatment of drug addiction which we feel is scientific and pragmatic. Obstacles to the realization of this project are the current legislation and cultural biases. What we can hope for is that educating the public will help correct the misconceptions that regard the problem of drug dependence and that the experience of other nations in the treatment of addiction may serve as an example for us.

## References


1. Cooper J R, Altman F, Brown B S, Czechowicz D, (Editors) . Research on the treatment of narcotic addiction. State of the Art. Treatment Research Monograph Series, N.I.D.A, Rockville, Maryland, 1983.
2. Maremmani I, Nardini R, Daini L, Zolesi O, Castrogiovanni P. Il trattamento chemioterapico della dipendenza da oppiacei con agonisti. Stato dell'arte. Quaderni Italiani di Psichiatria XI (3):233-264, 1992.
3. Tagliamonte A, Maremmani I, Castrogiovanni P. Disturbi da uso di sostanze. Il problema delle tossicodipendenze. In: Trattato Italiano di Psichiatria. Volume 2 (43), edited by Cassano G B, D'Errico A, Pancheri P, Pavan L, Pazzagli A, Ravizza L, Rossi R, Smeraldi E, Volterra V. Masson, Milano, 1992, pp 1136-1141.
4. Maremmani I, Castrogiovanni P. Disturbi da uso di sostanze. Disturbi da Oppiacei ed Analgesici. In: Trattato Italiano di Psichiatria. Volume 2 (43), edited by Cassano G B, D'Errico A, Pancheri P, Pavan L, Pazzagli A, Ravizza L, Rossi R, Smeraldi E, Volterra V. Masson, Milano, 1992, pp 1148-1161.

5. Maremmani I, Castrogiovanni P. Manuale del Trattamento Ambulatoriale con Metadone, SIMS, CN-MPC, Pisa, 1992.
6. Parrino M W. State Methadone Maintenance Treatment Guidelines, SAMHSA, DHHS, Rockville, MD, 1992.
7. Nardini R, Maremmani I, Zolesi O, Castrogiovanni P. Breaking through a wall of prejudices. In: Postersession, AMTA, Orlando, Florida, USA, 1992, pp 8-11-Novembre.
8. Cirillo M, Maremmani I, Nardini R. Il diavolo non esiste. Per un approccio diverso al problema della tossicodipendenza, Assessorato Cultura, Pietrasanta, 1984.
9. Maremmani I, Cirillo M, Nardini R. Etiopatogenesi della Tossicodipendenza da eroina: implicazioni per la prevenzione e la terapia. In: La tossicodipendenza da eroina fra progresso scientifico e pregiudizio culturale, edited by Maremmani I, Castrogiovanni P. Grasso Editori, Bologna, 1990, pp 13-32.
10. Damiani M S, Angelini G, Guareschi M, Maremmani I. Personalità dell'eroinomane e trattamento chemioterapico a lungo termine. In: La tossicodipendenza da eroina fra progresso scientifico e pregiudizio culturale, edited by Maremmani I, Castrogiovanni P. Grasso Editori, Bologna, 1990, pp 69-79.
11. Szasz T. The Myth of Mental Illness, Hoeber- Harper, New York, 1961.
12. Szasz T. The discovery of drug addiction. In: Ceremonial Chemistry, Anchor Press/ Doubleday, New York, 1974, pp 3-18.
13. Maremmani I, Nardini R. I trattamenti farmacologici sostitutivi come strumento in grado di ridurre lo spaccio e il relativo reclutamento di nuovi tossicodipendenti. In: Relazione a "La prevenzione delle tossicodipendenze sul versante degli interventi sanitari", Regione Lombardia, Milano, 1992,
14. Newman R G. Bridging the troubled waters. Audio cassettes of: Sixth Annual North-East Regional Methadone Conference (New Port, December 1989), Satellite Broadcasting, Frederick, MD, 1989.
15. Castrogiovanni P, Maremmani I. Atteggiamento psicoterapico e rapporto con il tossicodipendente in trattamento. In: La tossicodipendenza da eroina fra progresso scientifico e pregiudizio culturale, edited by Maremmani I, Castrogiovanni P. Grasso Editori, Bologna, 1990, pp 81-99.
16. Castrogiovanni P, Maremmani I. Tossicodipendenza e psicopatologia. Quaderni Italiani di Psichiatria 8:481-505, 1989.

*Springer-Verlag*
*and the Environment*

WE AT SPRINGER-VERLAG FIRMLY BELIEVE THAT AN international science publisher has a special obligation to the environment, and our corporate policies consistently reflect this conviction.

WE ALSO EXPECT OUR BUSINESS PARTNERS – PRINTERS, paper mills, packaging manufacturers, etc. – to commit themselves to using environmentally friendly materials and production processes.

THE PAPER IN THIS BOOK IS MADE FROM NO-CHLORINE pulp and is acid free, in conformance with international standards for paper permanency.

GPSR Compliance

*The European Union's (EU) General Product Safety Regulation (GPSR) is a set of rules that requires consumer products to be safe and our obligations to ensure this.*

*If you have any concerns about our products, you can contact us on ProductSafety@springernature.com*

In case Publisher is established outside the EU, the EU authorized representative is:

Springer Nature Customer Service Center GmbH
Europaplatz 3
69115 Heidelberg, Germany

**Batch number: 10745813**

Printed by Printforce, the Netherlands